A Beginner Beekeeper's Beekeeping Book

Massive Money Strategies, Beekeeping Supplies & Business Plan & Funding for Your Honey Bee Business!

By Brian Shawn

Table of Contents

DEDICATION

**This book is dedicated to my Father
Ulester L. Mahoney Sr.**

ACKNOWLEDGMENTS

I WOULD LIKE TO ACKNOWLEDGE ALL THE HARD WORK OF THE MEN AND WOMEN OF THE UNITED STATES MILITARY, WHO RISK THEIR LIVES ON A DAILY BASIS, TO MAKE THE WORLD A SAFER PLACE.

Disclaimer Notice

This book was written as a guide and for information, educational and entertainment purposes only. No warranties of any kind are expressed or implied.

Readers acknowledge that the author is not engaging in the rendering of legal, financial, medical or professional advice, and the information in this book is not meant to take the place of any professional advice. If advice is needed in any of these fields, you are advised to seek the services of a professional.

While the author has attempted to make the information in this book as accurate as possible, no guarantee is given as to the accuracy or currency of any individual item. Laws and procedures related to business, health and well being are constantly changing.

Therefore, in no event shall the author of this book be liable for any special, indirect, or consequential damages or any damages whatsoever in connection with the use of the information herein provided.

Introduction

INTRODUCTION

I'm not sure how old I was, but it was so long ago that I have no clear memory. The only reason I know about it because my parents taught me about it.

My grandfather was a commercial beekeeper, and he was one of the pioneers in this profession. He was an orchardist or a keeper of fruit trees, and he discovered that having beehives in his orchards improved the trees' productivity significantly.
He states "that any rise in productivity will make a significant change in income for any family." Which is amazing, isn't it?

Do you know that honey products, unlike many other goods like crops and cattle, have amazing business prospects and produce healthy foods?

Thinking about having your hives is not as difficult as it may sound. In fact, it has many benefits, including the collection of honey from your own garden or the knowledge that you're bees help in the pollination of local plants.

But beekeeping, on the other hand, is not an easy practice. Several important problems faced by people are lack of beekeeping expertise, how to start a colony, honey bee pests, and insufficient research about honey bee products.

So if you're new to beekeeping or want to get into this as a rewarding hobby or a money making business, having knowledge about beekeeping issues will help you prepare for any problem in advance, and help you with the harvesting of your hive.

INTRODUCTION

It is also possible that as a new beekeeper, you get scared by the initial cost of beekeeping. Keep in mind that if you are thinking about starting this business or doing it for fun, you'll need a hive, proper protective gear, a smoker, and a hive tool, among several other things. Therefore it is important to learn about it briefly in advance.

Diseases and pests are other very common and popular problems bees face. Therefore it is very important to learn how to keep your bees safe and how to solve issues that beekeepers like you in this field can encounter.

So if you are looking for any handy book, then you have come to the right place. Because this book has every piece of information that you need to know in order to start your business. It will genuinely answer some of your questions and provide you with a deeper explanation of the methods used.

It will also appeal to those who have never kept bees and have plans to do so. Even if you have little interest in bees, you can find it an interesting read.

In the first chapter, you will read about why honey is important and what are the different types of honey bees. Also, you will have an idea of how they work to collect honey. In the second chapter you will read about different honey bee species and how you can start a beekeeping business and what tools do you need to make massive money with all the several types of bee products!

INTRODUCTION

In the third chapter you will read about how to place your colony, with easy step-by-step instructions. In the fourth chapter I discuss different beehive pest problems and their solutions. In fifth chapter you will read about honey products and in sixth chapter will discover how to produce them. The seventh chapter is about bee safety, and taking the sting out of being a beekeeper.

This book also provides solutions to a number of beehive pest and health issues that you are going to experience. Its clarity and thorough explanations on all facets of beekeeping make it an excellent resource for every beekeeper.

All the information present in this book will teach you beekeeping and will introduce you to this beautiful world, so you can be happy and successful in your beekeeping endeavor.

I have studied university Entomology professors and have done a great deal of research on honey bees in past years and have deep knowledge about beekeeping and honey bees. The book will share my vision and expertise accrued over this long journey.

You will get amazing knowledge and a thorough understanding of bees and how they work. This book, will help to educate about all the important points of beekeeping, with some very valuable information for beekeepers to take note of.

INTRODUCTION

This book talks to you in easy to understand language and avoids using any technical jargon that would be difficult to understand for a beginner beekeeper.

Beekeeping can be both demanding and satisfying. So this book will assist you in facing any challenge that you may face in this amazing journey.

So get excited! You are about to take the first step into the wonderful world of BeeKeeping!

Chapter 1:
The Amazing
Honey Bee

THE AMAZING HONEY BEE

Do you know that honey bees play a very important role in flower pollination? You will easily distinguish them by their black and gold stripes, beautiful transparent wings, and signature body.

You will also find them flying from flower to flower, and it's as difficult to imagine this world without them, as it's difficult to imagine a world without trees. This hardworking winged creature has been pollinating our plants and creating the golden syrup which we call honey, for millions of years.

Bees are perfectly suitable for pollination, support plants in their development, reproduction, and food production. They normally perform this function by moving pollen from one flowering plant to another, thus keeping the life cycle going. According to research, pollination is needed for the distribution and survival of at least 30% of the world's crops and 90% of all plants.

The first step

So if you are thinking about becoming a beekeeper, then learning as much as you can about bees will be the first step toward becoming a good beekeeper. There will be chances that you might see something new every time you get into your hive because of several variables that can affect your honeybees.

THE AMAZING HONEY BEE

As a beekeeper, you must be creative in your ability to find out why bees behave in a certain way and how such behaviors can affect their well-being in order to make effective management decisions.

The European honeybee, Apis mellifera, is the most common bee species in the United States. There are more than 20,000 recognized bee species in the world. There are 4,400 different species of bees in North America alone, including social bumblebee colonies, solitary tunnel nesting bees, and solitary ground-nesting bees.

Honey bees devote their whole life to supporting the hive. Every bee has a unique job to do, and that job must be completed in order for the colony to survive.

Types of honey bees in a colony

Honey bees are amazing social insects who use a caste system to complete important activities that guarantee the colony's existence. For example, thousands of sterile female worker bees will do tasks like cooking, washing, breastfeeding, and protecting the colony. Male drones exist exclusively to mate with the queen, the colony's only fertile female.

Queen bee

Queen bee is normally recognized by her belly, which is elongated, can be seen through her folded wings. A Queen bee's average size (length) is 0.8 inches/2 cm.

THE AMAZING HONEY BEE

Queen bee's role

The queen's main role in the hive is most important as she lay eggs. She is typically the colony's only reproductive female. At the start of spring, when the first pollen is taken home by the workers, the process of egg-laying starts. The production of eggs will continue until autumn, or if pollen is no longer viable. Early in her life, she mates, stores millions of sperm in her body, and fertilizes her eggs as desired. She has the unique ability to lay up to 2,000 eggs in a single day.

She lays unfertilized eggs to produce male drones. A queen bee usually lives for three to five years. When she finishes laying eggs, the hive starts searching for a substitute and feed royal jelly to a developing larva.

The second thing she performs is the queen bee constantly releasing pheromones, a form of bee scent that can only be detected by the bees in the hive. These pheromones keep female workers germ-free while still serving as a warning to the rest of the colony that the queen is still alive and doing well. This different odor is thought to give the colony a sense of belonging and unique personality.

The lifespan of a Queen bee

It is said that a productive queen who is loved by the colony and has no disease can live for at least two years but may live for three or four years or even longer.

THE AMAZING HONEY BEE

Drones

A drone bee is another important part of a bee colony, and a male bee is born from an unfertilized egg. Drones lack stingers and have beautiful bigger eyes. They play no role in the protection of the hive, as they lack the body parts needed to gather pollen or nectar, so they can't contribute to the community's food supply. A drone bee's average size (length) is 0.6 inches/1.5 cm.

Drone bee's role

The main role of the drone is to mate with the queen. This mating process takes place in flight, which clearly explains why the drones need better vision, which is provided by their wide eyes. If one drone successfully mates, he will die shortly after sexual intercourse because the penis and related abdominal tissues will be torn from his body.

In the fall, worker bees keep an eye on the food supplies and keep drones out of the hive when they are not important, essentially starving them to death.

The lifespan of a drone

The drone's average lifetime is around 55 days. Drones who mate with new honey bee queens die almost instantly. However, drones have been reported to live for up to 90 days or 12 to 13 weeks.

THE AMAZING HONEY BEE

Worker Bees

Worker bees are the smallest part of the bee caste, but at the same time, they are the most abundant. Many of these workers are female, and, in most cases, they are unable to reproduce. These workers are unable to mate, but in a colony, if there is no queen, they will begin to lay unfertilized eggs, which grow into drones.

These workers bees have three clear eyes (ocelli) at the vertex and well-developed compound eyes on the sides of their heads. For sucking nectar from flowers, their tongue is well shaped and elongated. A worker bee's average size (length) is 0.4 inches/1 cm.

Worker Bee's role

These are called worker bees for a true reason: as they're the hardest working creature of this community! Except for laying eggs, these bees perform all of the critical tasks in a hive. Their age would decide which job they are assigned.

These workers perform all of the mandatory tasks in a colony, including:

- Hiding the hive's wax and forming it into honeycombs.

- They gather all of the nectar and pollen and took them into the hive, and turn it into honey.

THE AMAZING HONEY BEE

- They make royal jelly for the queen and the young larvae to eat.

- They also look after the larvae and queens' important needs.

- They cap mature larvae cells for pupation and clean the hive of wreckage and any dead bees.

These bees also protect the hive from burglars and keep it warm, cool, and ventilated to maintain optimum conditions in the hive. In the winter, their main goal is to keep the queen alive and warm by grouping around her.

The cluster gets more compact as the temperature decreases. Shivering worker bees generate heat, and they frequently switch from side to side between the inner and outer parts of the cluster. In this way, except in very extreme cold temperatures, no bee can freeze.

The lifespan of a Worker Bee

In the normal season, the worker bees live for around 38 days, and live 60 days in the spring, and even longer in the winter.

The life cycle of the Honeybee

The honey bee has four life stages: egg, larval, pupal, and adult stages of the honey bee.

THE AMAZING HONEY BEE

Phase 1 – The Egg:

Now you know that only bee queen in the colony is capable of laying around 2,000 to 3,000 eggs in a single day. On the third day, the egg is standing straight and has fallen to the side. It is important to note that a queen bee lay both fertilized and unfertilized eggs. So female bees or queen bees will emerge from these fertilized eggs. Male bees, who are also known as drone bees, will be born after unfertilized egg hatches.

Phase 2 – The Larval:

Three days after the egg turns into larvae and six days after the egg is laid in the beehive, the difference between a worker and a queen bee can be visible. During their first three days as larvae, all larvae, including female bees, workers, and drone bees, are fed the "royal jelly." During this stage, the larva sheds its skin several times. The royal jelly is later fed only to the female larvae, who grow into queen bees. Finally, the worker bees apply beeswax to the top of the cell to secure and support the larvae's development into a pupa.

Phase 3 – The Pupal:

In this phase, the bee will have grown wings, eyes, legs, and small body hair, and they will visually resemble an adult bee.

THE AMAZING HONEY BEE

Phase 4 – The Adult:

The young adult bee will make its own way out of the closed-cell after the pupa has matured.
Irrespective of whether it's a worker or a drone, or queen, the time it takes for each to emerge from the egg cell can vary slightly.

- It normally takes 16 days for a queen bee to develop from an embryo to an adult.

- It normally takes around 22 days for a worker bee to reach full maturity.

- It normally takes 24 days for a drone bee to mature into an adult bee.

The process of making honey

Honey bees try to produce as much honey as possible during the whole summer month's period in order to strengthen the colony during their "offseason." Honey is used as a source of nutrition for young bees. Honey bees that are newly hatched eat nectar and pollen to ensure that they become strong and start a job as spring arrives.

However, honey processing is a multi-step process, as you would expect. Let's find out how the honey bees make this valuable food for the colony one move at a time.

THE AMAZING HONEY BEE

Stage 1: Collecting nectar

In the first stage, when a worker bee sees a strong nectar supply, she starts the work! She sucks nectar from the inside of flowers with her proboscis (a long, tube-shaped tongue), and on a single searching trip, she can visit more than 100 flowers.

The nectar is usually kept in a special sac called a honey gut, along with a small amount of honey bee saliva. The worker bee will only return to the hive to empty the honey stomach and to drop off the load.

Stage 2: Passing the nectar to house bees

In the second stage, house bees, which live in the hive, wait for the hunters to arrive. The searcher bees will pass the nectar to house bees, who then begin the honey-making process. Enzymes alter the Ph. and other chemical properties of nectar when it is chewed and transferred from bee to bee.

The nectar and enzyme mixture contains excessive water to be preserved through the winter at this stage. It's up to the bees to dry this water.

Stage 3: Water evaporation from the honey

In the third stage, the worker bee will evaporate the water. As the honey is transferred from bee to bee, some of the water is lost. Bees, on the other hand, use two different ways to dry out their honey. In one way, the honey will be scattered over the honeycomb.

THE AMAZING HONEY BEE

This method expands the surface area of the water, allowing for further evaporation.

To raise airflow and evaporate still more liquid, bees may fan their wings above the honey. The water content of the honey would eventually drop to about 17-20%, down from a whopping 70%. The bees put in a lot of effort to get their food which is very interesting and commendable!

Stage 4: Storing the honey

Storage is the final stage in the whole honey-making process. The honey will be placed in the honeycomb cages, where it will remain until the bees want to consume it. Each cell is covered with beeswax to keep the honey fresh.

Beeswax

The hive's base is made of beeswax. Workers bees use Apis honeybees to build the hexagon-shaped honeycomb in which they live, work, raise their young, and store their food supplies. Beeswax is transparent in its normal state. When pollen or propolis stains it, it turns the bright golden color we equate with beeswax. Around 300 different substances make up beeswax. Its composition varies somewhat depending on the location of the honeybees.

THE AMAZING HONEY BEE

Propolis

This is a kind of paste made by bees to seal their hive and protect it from diseases. Propolis is an antibacterial, antifungal, and antiviral substance made up of beeswax, honey, and tree resins. It cleans and safeguards their hive. It's also really sticky, which honeybees love to use to plug any gaps or holes they come across when on a housekeeping trip.

That is how bees produce honey. However, they work very hard because, at one time, they will produce only one part of honey. One teaspoonful requires the work of at least eight bees over the course of their lives. Luckily, they normally produce more than they want, allowing us to get some as well.

Honey Bee Dance

As honeybees are unable to communicate verbally, they use special dances for communication. They will use dances to give a variety of signals, ranging from the need to swarm to the specific location and distance to a food source.

The circle dance and the tail-wagging dance are the two forms of bee dances, with the sickle dance serving as a transitional style. The energy of the dances is determined by the consistency and quantity of the food supply in all situations.

THE AMAZING HONEY BEE

Round dance

Round dance, as the name suggests, is a circular movement. This specifies that the food supply is within 50 meters of the nest.

Waggle dance

Field bees, in fact, perform the "waggle dance." When they discover an abundant supply of nectar, they return to the hive and perform the dance to alert other bees about the location of the flowers. The waggle dance is a figure eight-pattern performed by a bee as its abdomen waggles, and it is used to find food that is more than 150 meters away.

The dance also represents the position of the flowers in relation to the light, and the bees change the dance automatically as the sun travels through the sky. The length of the dance may be used to communicate the exact distance. A longer dance denotes a considerable distance.

As you can see, from their ability to communicate, their social organization, and the valuable work that they do, the Honey Bee is truly amazing!

CHAPTER 2:
Types of Bees &
Beekeeping
Supplies

TYPES OF BEES & BEEKEEPING SUPPLIES

Beekeeping isn't a kind of hobby you will select instantly. Unlike jogging or knitting, you must first have the necessary equipment before you can start. If you're a novice, there are two things to think about before you begin.

First, you need to decide which breed of the bee to order, from whom to order, and secondly where to procure packages and queens, etc. Let's start with basic tools and equipment.

To get you started, here's a list of important beekeeping supplies:

Bee Hives

Mostly, new beekeepers will prefer to buy ready to assemble hive parts, but building your own hive is definitely possible. If you do, it's important that you stick to the exact dimensions for the hive that you like. Sometimes a honeycomb is built where it is not needed due to incorrect hive dimensions.

You must have a clear idea of where your bees will live when you build them. Beehives are divided into three categories.

•Langstroth Hive

•Top bar hive

•The Warré Hive

TYPES OF BEES & BEEKEEPING SUPPLIES

Let's find out which style is best for you.

Langstroth Hive

Reverend Lorenzo Lorraine Langstroth (1810-1895), a native of Philadelphia, created the style in 1851. This structure is made up of square boxes stacked on top of one another with interchangeable frames for the bees to create comb in.

Parts of a Langstroth Hive

Here are important parts of the Langstroth hive.

Top/Vented Cover:

This part of the hive keeps it dry in the event of a rainstorm. It's equal to a house's roof.

Inner protective Cover:

The inner cover is found in between the top hive box and the outer cover. It protects the frame and keeps it from sticking to the outer shell. During the process of honey extraction, it can be used for a bee escape.

Hive Frames:

Removable frames are found in the hive boxes. Frames are available in different sizes. Bees use the beeswax foundation as a guide to creating honeycomb within the wooden frames. Young bees, pollen, nectar, and honey are stored in different comb cells.

TYPES OF BEES & BEEKEEPING SUPPLIES

Queen Excluder sheet:

This part of the hive only allowed worker bees to fly in, preventing the queen and drones from accessing the honey. This is a piece of optional equipment that keeps the queen from laying eggs in the honey supers. However, an excluder isn't used by any beekeeper.

Foundation:

It is important to keep in mind that within the boxes, most beekeepers use sheets of beeswax (or plastic) base as a guide. This encourages the bees to build straight comb within the plates.

Shallow Super:

The most common size for honey production is shallow supers.

Deep super or Brood box:

It's also known as the Brood Chamber, and it has more frames than the shallow mega. The queen will lay eggs for the next generation of bees in this part of the hive. Nurse bees look after the babies in this maternity ward.

The base of the hive:

Bottom boards are available on either a strong or screened bottom.

TYPES OF BEES & BEEKEEPING SUPPLIES

Any variation of the three super box sizes: brood, mediums, and shallows can be used in a Langstroth hive.

Top bar Beehives:

Top bar beehives are also a very good option for buying. Especially backyard beekeepers and organic growers are increasingly turning to top-bar hives. It is the world's oldest and most widely used type of beehive. A unique collection of horizontal bars is placed over a trough-shaped hive covered by a hinged or removable cover, and bees naturally create their comb downward from these bars.

You will find no doors, and there is also no need for a floor to hold the hive level. Simple wooden wedges or strips slip into slots to ensure that the bars hang straight. A top-bar hive is quite simple to build, but commercial top-bar hives are also available in the market.

The brooding area for the bees is defined by a divider board that confines the first 8 to 10 bars adjacent to the hive's opening, where the bees enter and leave easily, in more intricate versions of this style. The divider board is pushed laterally, and more bars are inserted as the colony expands and comb and honey fill the bars. When the bars are wrapped in a honey-filled comb, harvesting is as simple as lifting them out.

TYPES OF BEES & BEEKEEPING SUPPLIES

Warré Beehive:

Another amazing bar style is Émile Warré's (war-RAY) hive, which he produced in the mid-20th century. This special hive is referred to as a vertical top bar hive rather than a long horizontal top bar hive. You will find no frames or base sheets in identically spaced stacked boxes. Honeycomb is built by bees from top bars inside each box.

Empty boxes are usually set at the bottom of the stack rather than on top to allow the colony more overhead space. Beekeepers who use the Warré style sometimes "bottom-super" their hive. They believe that this arrangement more closely resembles bee life in the wild.

So for the first few years, I will recommend that new beekeepers should start with a Langstroth Hive. These, in my opinion, are the best kinds of beehives for beginners because they still have a lot to learn about beekeeping.

Point to note: Whichever option you want to choose to start your hobby, it's a good idea to start small, so you don't lose money or time if you plan to switch methods later.

Smoker

Do you know that smoke, when used properly and in moderation, can help to relax bees? For a pair of hives, a small smoker would serve the purpose.

TYPES OF BEES & BEEKEEPING SUPPLIES

You will want a bigger one if you have four or more hives in your beehive. The aim is to create cool white smoke. You should buy smoker fuel or use dried pine needles in your smoker to do so.

Bee Suit

Your bee suit would be an important investment. You may purchase less expensive suits, which is perfectly acceptable when starting out any new business. However, you need to keep in mind that some of the higher-quality suits will be more useful in the future.

Hive tool

It's a kind of flattened iron Spatula tool. Its one end sharpened for inserting between hive boxes to separate them, and the other bent at 90 degrees to separate the frames. You will used it to scrape off excess comb or debris from various parts of the hive, as well as bee glue (Propolis).

Bee Brush

In Apis mellifera colonies, a bee brush is typically used to brush off the bees from a honeycomb before it is removed for extraction. It can also be used to bring the scattered bees together during a swarm hiving.

TYPES OF BEES & BEEKEEPING SUPPLIES

Bee veil or Bee hat

It is used for face protection from bee stings. It is made of black mosquito nylon netting with cloth on top and bottom. The bottom cloth should have an elastic rim to keep it in place around the neck.

Hand Gloves

They're typically made of heavy canvas, supple leather, or rubberized cloth and great for building confidence in beginners. Its wrist has an elastic arrangement that will shield the hands from bee stings.

Shoes

Shoes are very helpful gear, especially when a bee can crawls up your pant leg and stings you. This is one of the most painful stings when it occurs unexpectedly and in a sensitive place, such as your thigh.

Queen catcher

The queen catcher is also very useful, especially when you are going to remove the queen for a short period of time. This handy tool allows you to trap and inspect your queen without causing any harm to her.

TYPES OF BEES & BEEKEEPING SUPPLIES

Honey extractor

It's a kind of hand- or motor-driven machine with a rotating chamber into which the frames match, and it extracts honey in its purest form from the honeycomb. By using its centrifugal energy, honey bubbles out of the frames without splitting the comb.

Honey Bee Stock

You'll definitely need to buy the first bees until you capture a swarm. Therefore it is important to know what types of bees are available before you can buy the first package. This will also assist you in determining which bees will succeed in your area. Point to note: It's important to order bees in advance so you can get them home in April or May.

It's important to note that there are no "natural" bee species because your bees can mate with other bees in the field. This will result in a different bee species mixture.

The Italian Bees

The Italian worker bees are light-colored, while the queen is a little darker, making her easier to see. The abdomens of worker bees have overlapping stripes as well.

Italian bees, which originated on Italy's Apennine Peninsula, were introduced to America in 1859. It displaced the black or German bees brought over by the first colonists.

TYPES OF BEES & BEEKEEPING SUPPLIES

The most common bees to order are Italian bees. They're famous for being very gentle and making a lot of honey. They are usually raised in the south and can easily live in colder temperatures because they need more food to pay for their inability to form a tight cluster like other honey bee forms. Italian bees are excellent hunters and maintain their hive in excellent condition.

However, if you want to know some of their drawbacks, these Italian bees swarm, and their sense of direction isn't as good as other bees', so they can wander from colony to colony and steal frequently. This can lead to disease transmission between hives.

The Russian Bee

Russian bees are found in two colors: dark brown and black, with a paler yellow belly. In 1997, the Russian bee was introduced to the United States. Bees in the United States were suffering from very serious issues like colony failure that caused some beekeepers to destroy up to 90% of their hives.

The USDA responded by introducing the Russian bee, which was noted for its parasite resistance. They hoped that this stock would help save US bees because parasites were one of the reasons linked to colony collapse. These bees are a little fiercer than Italian bees.

TYPES OF BEES & BEEKEEPING SUPPLIES

The Buckfast Bee

These types of bees are found in yellow and brown in color and are like honey bees in their appearance. Buckfast bees are a hybrid. Brother Adam of Buckfast Abbey in southwest England developed them in the twentieth century. The stock was brought to the United States through Canada and is now widely available.

Buckfast bees are resistant to Tracheal mite and can prosper in cold climates. They are calm, simple to work with, and contain a lot of honey. They have a low tendency for swarming and are economical in their winter markets.

Buckfast bees are used to living in cold, rainy winters, so they're used to rapidly expanding their hive size in the spring. These bees are a good choice for backyard hobbyists who want to get some honey out of the bargain.

The Carniolan Bee

These bees are also very popular honey bee species. Carniolan bees have a dark belly with brown spots or amazing patches. You will find them a little smaller than most bee races, but that doesn't seem to affect their ability to search and return to the hive with pollen and nectar stocks.

TYPES OF BEES & BEEKEEPING SUPPLIES

The lazy nature of these bees is famous, but they are also considered among heavy spring producers. They still have one of the longest tongues of all bee species, making the pollination of crops like clover simpler.

This bee strain is more tolerant of cooler temperatures and can overwinter successfully. One of the least enticing aspects of these bees is that they like to swarm than other species.

Purchasing your bees

As a novice, it's better to start with buying bees. It is also the simplest way for an inexperienced beekeeper to launch a beehive. The package bees or a nucleus hive are the two most common methods of receiving bees. Let's find out about them.

Package Bees:

First of all, you will need to contact a local beekeeper supplier or a local beekeeping society to order a box of bees. A queen, some workers, and a sugar syrup feeder are included in most sets. Installing the kit bees in their new home and adding the queen bee to the staff should be covered by the bee supplier. She is kept secure inside a special cage that is included with your box of bees.

The indirect tactic is the most traditional way of adding a queen. As they will slowly eat their way through the food plug in her cage, the worker bees get to know the new queen.

TYPES OF BEES & BEEKEEPING SUPPLIES

Nucleus Hive:

You can also order this hive. A nucleus (also known as a "nuc") is known as a half-colony. A 5 frame nuc is the most common size. You'll get 5 comb frames, bees, pollen, a queen, and a brood (baby bees). When you buy a nuc, you get a head start on colony formation.

However, this method is dangerous than using package bees since the honeycomb from the donor hive will spread pests and disease to your hive. You need to find out where to buy good bees in your area by contacting a local beekeeping association.

Now you know all about the different types of bee hives, the different types of bees you have to chose from, and their characteristics. As well as the basic equipment needed to get started!

Great stuff! And we are only in the 2nd chapter!

CHAPTER 3: Planning Your Colony

PLANNING YOUR COLONY

Planning your colony is essential to having a productive healthy beehive: firstly, you will need to learn the rules to start this beekeeping; secondly, you will choose the best site for your hives and ensure that you have the proper habitat.

Bees only need a dry hole, food (nectar, pollen, or syrup and pollen substitute if natural supplies are unavailable), and parasite control. My simple rule is that you do not do something until you fully understand what you are doing! Therefore I have tried to cover everything that you need to know in order to start this business.

Here's a step-by-step plan to guide you and to answer any questions.

Step 1: Choose the location

First of all most important thing about bees is that they can survive in almost any climate. So it is important to choose a location first. Try keeping the hive under the rising sun (East) or toward the (South). The sun on the hive in the morning is helpful because it warms the hive and the bees. It is also important to shield them from the dampness; they must be kept off the ground. You can also pour cement to make maintenance easier after arranging the field.

Always place your hives in a safe location. Hilltops should be avoided because they are always windy. Also, stay away from low spots that retain cool air for prolonged periods of time and are humid.

PLANNING YOUR COLONY

Make sure your hive area is clean and has no water issue so you can easily get to the apiary.

Bees require water any day of the year, in addition to sunshine. So make sure to have water nearby. If you or your neighbors have a bathing pool, set up an alternate water supply for the bees when they visit so that the chemically treated water would not tempt them. The swimming pools should be closer to the hive than this source.

Bees want nectar and pollen. Will you have to feed the bees in order to keep them alive? Yes, you may need to assist the bees with feeding at times. This is particularly true during droughts or to assist a weakened hive in preparing for the winter. This leads us to the topic of food sources (we will discuss this issue in a later part of this chapter.)

Finally, bees want privacy. Place the hives away from high-traffic areas, such as playgrounds, swimming pools, and pet areas. Place beehive at a place that is ideally 50 feet away from high-traffic areas, but if space is small, put the hive near a tall fence or hedge. Bees and humans will always be happier if they are away from noise.

Step 2: Installing the nuc/package

A nuc or bee package includes a healthy young queen and a beautiful cluster of bees. Since a nuc already has a family, it can grow even faster. Which is nice when you're new to beekeeping,

PLANNING YOUR COLONY

I will recommend you to start with a small nuc to give you more time to learn before it develops into a massive, protective colony.

How to transfer the nucleus colony?

After picking up your nuc box, you need to go straight to your beehive to avoid any overheating in the car.

Then place the nuc box on top of your hive box or in it as you like. Then, on one of the sides near the bottom, open the entrance to the nuc box. The bees will be most likely to come out but don't worry; soon, they'll quiet and get to work fast.

Also, you don't need to light your smoker for this part, but it's a good idea to use your new bee veil and hive tool as the test. Your hive tool will assist you in cutting the tape and opening the door if the supplier has taped its entrance.

Since the bees will be placing themselves in their new home, you will see a lot of noise in the hive. Pollen and nectar collection starts almost immediately, so keep an eye on it as well.

You need to follow these steps.

•First of all, put on your beekeeper's cap, veil, or suit to get dressed properly. A nuc colony could be more resistant than a group of bees. They have a brood which they want to protect.

PLANNING YOUR COLONY

•Secondly, turn on your smoker. Puff a small amount of cool white smoke near the nuc's entrance. This will have little effect on the bees, but it does help to reduce the warning response.

•You need to prepare five frames for your empty hive. (Assuming you're using a Langstroth hive of 10 frames.) Place two frames on either side of the hive.

•Now pull out your nuc box and try to open it. Start this very carefully, extracting one of the frames outside of the nuc case. We have no idea exactly in which frame the queen is living. Do not crush or roll your queen!!

•After you've placed all of the frames, including the queen's frame, in the new hive, to complete your 10-frame formation, add the last empty frame to the hive.

•Now you will close the hive and install your feeder. It's very important to feed your nuc. Even though in the start, you have 5 drawn comb frames, they still have a lot of work ahead of them.

•Remember to stand properly your entrance reducer one more time before setting up a nucleus colony. Until the population breeds, we want the entrance to be a little small.

One week later: It's time to check your hive for the first time.

PLANNING YOUR COLONY

Point to note: It's somehow a difficult challenge to go from a nuc to a Warre or Top Bar. It is possible, but it is not a simple procedure and requires proper understanding. As a result, a new beekeeper can only accept a nuc if a Langstroth is their preferred hive.

How to transfer package colony?

1.   One thing that you need to keep in mind after receiving the box is to put it in a cold, dark place for some hours to encourage the bees to 'rest' before installing them into their hive. Be sure to save the bees from excessive heat or cold, as well as noisy sounds or unusual vibrations. At this stage, you can spray the bees with sugar syrup (1 part sugar, 1 part water) on a daily basis before you're able to place them in a hive.

2. Now it's a good time to double-check if all of the hive equipment that you have bought are in working order.

3. After you have set the equipment and you're able to place the bees in the hive, feed them sugar syrup again and take the box into the apiary (by holding the wooden sides). To stop being stung through the screen, keep your hands away from the screened sides of the box. Place the box in a shady area on the field.

4. Now remove four frames from the brood chamber's middle to make some area for the bees in the hive.

PLANNING YOUR COLONY

5. Now you will remove the wooden panel from the first box of bees with the hive tool.

6. Please very carefully extract the tin feeder and queen cage from the box's top hole.

7. For checking the queen, you need to shake bees from the outside of the queen cage to ensure she is still alive and working well. Place the queen cage in a shady location. To avoid bees from escaping, replace the wooden panel over the hole.

8. In the hive, put the queen cage.

9. Only before putting the bees in the hive, powerfully knock the packet on the ground once to push the bees to the bottom of the attachment. When doing so, check carefully to keep the wooden lid in place.

10. Now remove the wood panel and inverse the kit over the hive body easily. Shake the bees strongly into the hive's room. It's also expected that you'll need to shake the box many times. If there are a lot of bees buzzing around, no need to worry; they're just "confused" but will not be aggressive, and they'll calm down and join the hive soon.

11. Put the package in front of the hive's door, helping bees to crawl into the hive. After the workers have scattered on the bottom board, just return the frames to the hive, here you need to be very careful so as not to crush any bees.

PLANNING YOUR COLONY

Now it is the time to place the Queen in her castle?

1. For placing the queen in her castle with the white sugar candy, you will remove the plastic cap from the long side of the queen cage. You will see that within one or two days, the bees will eat the candy and release the queen.

2. This process will help the bees to become used to the queen, dropping the chances of the queen being rejected. Keep in mind that you will not drop the cork from the end of the bottle without the candy!

3. Now, place the queen cage candy side up between the hive's two center frames. Ensure that the cage is tightly secured between the frames, so it does not slip to the hive's floor.

4. Feed sugar syrup to the new colony. It is important that the bees have a constant supply of food; mainly, the colony has an adequate supply of honey stored.

5. Replace the inner and outer covers, as well as the lid.

Note: Don't forget to check the colony after the first 5 days that the kit is installed to make sure the queen is still alive and has been released. After another 5 days, check the colony carefully to see if the queen has started laying eggs. In the middle of cells, eggs emerge as tiny grains of rice standing up. At this point, if necessary, add more sugar syrup.

PLANNING YOUR COLONY

After you successfully install the bee Package

•One thing to make sure is not to touch the hive for at least a week. I know this is the hardest part. But it is a very important point. If you create disturbance for the bees, they will kill their queen instead of welcoming her.

•Make sure the feeder is loaded perfectly.

•You will see them flying everywhere. They're maybe puzzled. Soon, you'll see them doing orientation flights outside in front of the hive. This will last a few days before settling down.

•Keep an eye over bees as they are taking dead bees, also known as undertaker bees. This is one of the first things you can see, and it will inform you that the colony is doing well.

•Keep an eye out for bees carrying pollen on their wings, which they use to feed their brood. This is an indication that you will soon have a laying queen.

Advantages of Nuc Over a Package

It is true that the colony is already generated.

•Now the workers and drones are acquainted with the queen, so there is no need for a new "acclimation" process.

PLANNING YOUR COLONY

•The colony has all kinds of bees ranging from every stage, from freshly laid eggs to adult bees.

•The nuc now produces honey and pollen, all of which are necessary for the colony's existence.

•The colony is active and able to continue hunting right away.

•When dealing with a nuc, you use the bees and frames that came with it. As a result, you simply put the frames in a Langstroth box, and the process is complete.

Advantages of Package over a Nuc

From the box, the bees are now placed into the hive. You can have any new beehive like Langstroth, Top Bar, or Warre in style. You will be able to use any of these hives once bees are set in their hive.

Step 3: Feeding your bees

Now talk about the third step, which is feeding your bees.

Let's start with "WHY."

You need to check carefully that at least one of the outer combs contains preserved honey because it is possible that your nuc may arrive hungry.

PLANNING YOUR COLONY

There is no need to feed if there is a strong nectar flow when the nuc is attached, but feeding sugar syrup to the small colony frees the bees from the need to quest for nectar and allow them to focus their attention on the collection of pollen, rearing of brood, production of beeswax, and drawing out comb instead.

The processing of beeswax, which is used to "draw out" base frames into "drawn combs," requires a large amount of sugar (whether from nectar or syrup). As a result, feeding sugar syrup to the new colony on a daily basis before all of the frames in the lower box are entirely drawn is usually a good practice.

When and how to feed your bees?

As you know that bees can only create comb on warmed frames, they will not be able to draw comb outside of the cluster unless the weather is very hot. Overfeeding would cause the bees to fill the brood nest with nectar, stopping the queen from laying further eggs.

So it is important for you to keep a close eye on the brood nest and feed enough syrup for the bees to start producing "white wax" along the edges of their cluster, which means that the amount of nectar and syrup, relative to the amount of free open cells, is stimulating them to activate their wax glands.

PLANNING YOUR COLONY

When the nuc's enclosed brood arrives, the colony (cluster size) will increase, and you can feed heavily to enable them to attract comb.

How much sugar syrup?

In the start, the colony will only need up to a cup of syrup per day. Later on, almost a gallon per day will be needed. When all of the frames in the first brood chamber have been drawn, you will determine whether or not to continue the feeding process.

By that time, the main honey flow in the Sierra may have started, and you may want to pick some honey. However, you should still feed them again later in the season (in July) if they need it to help them draw out all the combs in the upper brood chamber and save some honey stocks for the coming winter.

Choose the correct sugar type for your bees

It is important to choose the right sugar. If you use the wrong sugar, the bees will become ill and suffer from poisoning or nosema. So, which sugar will you select? I have given some suggestions; please check out the list below.

Sugar types you can use:

•White or granulated sugar

•Cane or beet sugar

PLANNING YOUR COLONY

•Organic sugar

•Cubes of sugar

•Cane juice that has been evaporated

Organic sugar Vs. Cane sugar

Regardless of the fact that organic sugar can be eaten, a 2009 report found that organic sugar produces more ash than normal cane sugar (almost 0.03 percent ash in cane sugar vs. 0.20 percent ash in organic). While these figures will seem unimportant, bees can more freely eat the lower ash sugar. As a result, you should avoid using organic sugar.

Sugar types you cannot use:

•Brown sugar

•Turbinado sugar

•Confectioners' sugar

•Sugar from Demerara

•Palm sugar and coconut sugar

•Stevia

•Xylitol

PLANNING YOUR COLONY

- Sucralose

- Aspartame

- Corn syrup with a high fructose content

- Agave nectar

- Maple syrup

How to make homemade sugar syrup?

Usually, liquid feed supplementation is needed when the hive's honey supply is limited, such as in late winter or early spring. Most beekeepers use two simple sugar syrup recipes.

Spring season Feeding Sugar Syrup

It makes no difference whether you calculate by weight or volume. You need to add a percentage of dried sugar with a percentage of water. You may use cups or weight to compare (for example, 4 cups sugar to 4 cups water). Since it includes equal parts water and sugar, this ratio produces a 1:1 sugar water formula.

Fall season Feeding Sugar Syrup

For making sugar syrup for feeding, 2 cups of sugar and 1 cup of water will be used. Because in the fall, you will use a different recipe.

PLANNING YOUR COLONY

The syrup will have a double ratio of sugar to water. For instance, Eight-cups of sugar plus Four-cups of water.

Here is a recipe for your help.

Ingredients:

•Sugar

•Water

•Container

Instructions

•First of all, heat the water on stove to melt the sugar, the water does not need to boil; it just needs to warm up.

•Now add the sugar and whisk until the syrup cook well.

•Let the syrup cool before placing it in a feeder.

Note:

Add one drop of essential oils to avoid mold growth. It also provides a nutritional support to the bees. One thing to keep in mind is that this sugar ratio is just an approximate calculation. There's no reason to leave out anything if you have a little extra water or sugar than the same 1:1 or 2:1 ratio.

PLANNING YOUR COLONY

Timing of feed is important because bees won't eat syrup that's below 50° F (10°C).

Step 4: Inspection of the beehives

You must be thinking that beekeeping is just a process of simple observation and reaction. If you're a beginner beekeeper, you will need to check the hive once a week for a few months to learn the whole process. Change the schedule to every two weeks until you're ready.

Make sure the hive is neat and free of bee poop, the main landing board is litter-free, and there are no ants near the hive. Check the frames for larvae and eggs when you open the hives but on warm days only. You'll see a lot of larvae in different stages of development if the queen is working well.

However, you will need to consult with a specialist if you don't see any signs of a stable queen. A good place to start is the local beekeeping association. Finally, the hive's health will improve if you check it less often. To keep the bee's calm and relaxed when opening and thoroughly inspecting the hives, it's important to use smoke. The bees are stressed as a result of this, and it takes them about a day to heal.

When you get experience, you will discover that you don't need to pull several frames to find out what's going on inside. And just watching the bees as they come and go from the hive will help you a lot.

PLANNING YOUR COLONY

Step 5: Expand the hive when necessary

You can start this process by building one deep hive body-brood box. Cover it with a second brood box after the bees have packed it with 8 frames of bees and brood. Now let the bees build brood cells in the second brood box too. When the second brood box is fully populated (7 or 8 frames of bees), add a queen excluder, if desired, and then the honey mega the box that you will use to collect most of your honey.

Queen rearing

A new queen should be raised for a hive for three important reasons: emergency, swarming, and supersedure.

The emergency:

The emergency response occurs when the queen bee inside the hive is unexpectedly lost.

The swarm:

The swarm response occurs when a new queen is raised due to a shortage of space in the hive, and the hive can swarm as soon as the queen cells are capped and before the new virgin queens emerge from their queen cells.

PLANNING YOUR COLONY

The Supersedure:

A failing queen, either due to age, illness, or accident, prompts the supersedure reaction, in which a new queen is raised, and the failing queen is replaced.

Why Beekeepers Raise New Queens

An experienced beekeeper who knows why the colony is acting in a certain way will create conditions that allow the colony to produce new queens. There are many reasons why you should think about raising your own queens.

Replacing the Old Queens

Replacing the old queen with the new is a choice for beekeepers who observe low levels of new brood in their hives. This means that at peak nectar flow, the colony has enough workers hunting for nectar and pollen to keep the hive well supplied for the winter.

Splitting the bee Hives

Beekeepers can select to divide their large colony into two or more new hives to prevent swarming. Providing a new queen to one or both of the new colonies guarantees a constant supply of new brood.

PLANNING YOUR COLONY

Production Increased

A new queen produced from the larva of a highly productive colony will help a less productive colony in increasing its efficiency. The colony's genes control many characteristics such as disposition, disease and pest tolerance, seasonal population changes, and honey production rate. Staff from the current queen will soon replace the older population, and the colony will begin to exhibit the desired characteristics.

Why raise Queens?

You might want to re-queen your own hive each season so that you have a young and healthy queen. You can also breed new queens to sell to other beekeepers.

What you need to raise Queens

The grafting process, the non-grafting method, and the natural method are all viable methods for raising queens.

If you're considering grafting, you'll need to know whether to use the JZBZ system or the Beetek Bozi cell system for queen rearing. When grafting, the age of the larvae is very important; the optimal age is between 0 and 24 hours; any later and the graft will take, but the resultant queen will be inferior with less ovaries. Get a good queen rearing kit, that includes everything you'll need to get started grafting.

PLANNING YOUR COLONY

The Jenter and Nicot systems are non-grafting options to use if you are unsure about grafting the larvae and putting them in the cell cups.
The natural approach involves dividing your hive into two halves and encouraging the bees to raise the queen.

When is the best time to re-queen?

There are two types of seasons for raising queens, each with its own advantages and drawbacks. These seasons are Spring and Fall.

If you want a new queen in time for the honey flow, spring is the perfect time for queen rearing. Rearing in the spring also assists with swarm management and decreases the risk of theft.

However, due to the weather's unpredictability and the scarcity of drones, getting the virgin queens mated can be very difficult. Autumn has more reliable conditions, and you will normally found more drones around, making it easier to mate virgin queens.

Seasonal Hive Management for beginners

Seasonal hive management is also a very important skill to learn for beginners. You must ensure that your bees are ready to enter new seasons of their lives. Every season, there are a few things you'll need to do as part of beehive management to ensure that your hive thrives properly.

PLANNING YOUR COLONY

Spring

Your bees can have a difficult time in the spring. Warm and cold weather cycles can be unpredictable, and your hive is not working properly after a long, cold winter. However, if you've spent the previous months preparing for the coming months, everything is going to be perfect.

Your bees will be planning for the working season at this time of year. They will begin foraging once the weather becomes warm. As a result, they'll begin restocking their food supplies. Ensure that you continue to feed your bees until they are ready to work for themselves.

Tips for spring

•Make sure to check hives on hot days

•Always check your queen and her eggs

•As your hive expands, add supers and repair frames

•Feed supplementary food if essential.

Summer

Summer months can be humid, and bees have a long list of foraging tasks to complete in preparation for winter. In general, you should check a beehive every 1-2 weeks to ensure that everything is going well as part of beekeeping. You also maybe have to replace feed if there are droughts or other natural disasters.

PLANNING YOUR COLONY

Tips for summer

•Always check your hive after every week.

•Ensure that your bees have an adequate water supply nearby.

•Check queen and eggs production.

•Check the production of honey.

Autumn

Since bees are highly receptive to their surroundings, the queen will minimize her egg-laying activities when autumn comes. However, worker bees will continue to quest and work until the last possible moment in order to prepare for winter.

Tips for autumn

•Check the food supply carefully in your hive.

•You will check your hive every couple of weeks.

•Make sure that your hive has proper ventilation now.

Winter

If you have properly prepared your hive in the fall, you will have little work to do during the winter except to provide extra feed if necessary.

PLANNING YOUR COLONY

Bees will assemble for the winter, forming a close ball, which will only leave for food and to relieve themselves. Bees like to remain clean and tend to keep their surroundings neat.

Tips for winter

• Always add a feed-in winter as they may need it.

• Make sure if they are fine by tapping over the beehive to notice any buzzing.

• Take care of predators.

• Try to provide water nearby.

When planning your colony you have to pick a good location, be careful about water, select the proper sugar and alter your care as the seasons change.

In the beginning it might seem like a lot of work. However with pratice, it will soon become second nature, and you will get to enjoy all of the benefits of having a healthy productive beehive!

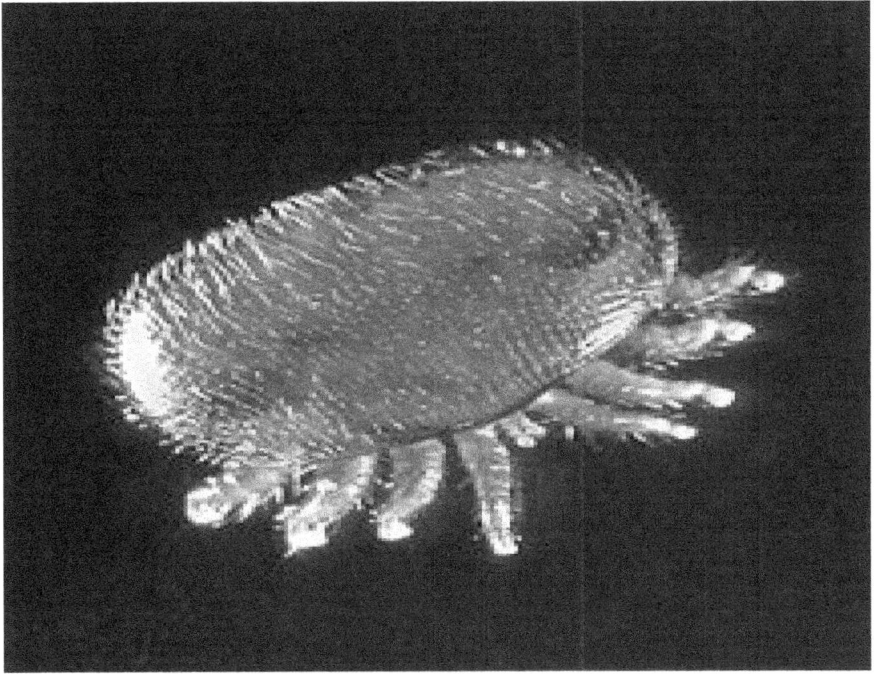

CHAPTER 4:
Pests & Bee Hive Health

PESTS & BEE HIVE HEALTH

Honey bee pests are a big challenge for hives all around the world. It can be very difficult for inexperienced beekeepers to identify what to look for and how to deal with pests and predators. This chapter is going to guide new beekeepers about the most common hive problems that they can face.

Pests

There are different types of pests. Some of them are as follows.

Varroa Mites

Varroa mites are one of the most dangerous threats to honey bees and beekeepers around the world. These mites are most difficult to deal with since they endanger the hive's existence after they have established themselves. They get their feed from adult bees' blood and lay their eggs in brood cells, where their larvae infect and consume bee babies by infecting them with viruses.

Varroa management requires early detection, close observation, timely treatment, and good beekeeping practices. Monitoring natural mite fall, drone brood screening, or the ether roll test can all be used to determine mite levels in the hive.

Close inspection needed

For close inspection of the mite, you will do a 24-hour count of natural mites to find out a clear estimation of the invasion level in a hive.

PESTS & BEE HIVE HEALTH

To catch the mites, you will coat the bottom of your Country Rube board in petroleum jelly or cooking oil, keep it into the lower part of the bottom board, will wait 24 hours, then take it out and count the mites. If you have more than ten mites per brood box, you have a problem in your hive.

Signs

Here are some important signs that show that your beehive is infected by them.

•Adult mites are visible.

•Bees often show varroa mite infestation with deformed wings, short abdomens, and deformed legs.

•You will also find a reduced adult population.

How to control?

Varroa control is very important for colony survival. To control the natural oils and acids, such as thymol and oxalic or formic acid, are used by the majority of beekeepers. There are many commercial formulations used all over the world. Beekeepers use a variety of techniques to save their hives.

Trapping the Varroa Mite

Varroa mites may also be caught through the use of drone frames.

PESTS & BEE HIVE HEALTH

Drone comb cells are bigger than worker comb cells, and these frames are specially designed to enable bees to make them.

The drone comb is an excellent mite trap because varroa mites favor drone brood 10 to 1. Replace the drone frame for the next cycle just before the drone's hatch (24 days after the eggs were laid), and destroy the drone comb. Our queens don't need the drones to reproduce because they've already mated and had a lifetime supply of sperm inside of them.

Apiguard

It is a Varroa mite remedy dependent on thymol, a part of botanical thyme oil. The maximum dosage of Apiguard consists of one foil pack inserted in a hive for around two weeks, followed by a second foil pack placed in the hive for extra two weeks.

Formic acid

Formic acid is more toxic than pickguard, destroys mites by gassing them. It made honey poisonous for humans, so it's only used in the fall and winter when the nectar flow is slow or non-existent. When using it, you must wear a respirator.

Tracheal Mites: Acarapis woodi

The honey bee tracheal mite is a second mite that invades honey bees. Within the thorax of adult honey bees, the tracheae, or breathing tubes, are home to this dangerous internal parasitic mite.

PESTS & BEE HIVE HEALTH

These mites can also be found in the thorax, abdomen, and head's air sacs. The mites use their mouth-parts to cut the breathing tube walls to feed on the bees' blood.

Signs

•You will see these bees outside the colony, where they will be crawling on the grass.

•They will crawl up the edges of the hive, collapse back down, and try again.

•The wings of the bee will not be working properly.

•The abdomen will look swell.

•In the final stages, bees will leave the hive.

•Give a report of bees in alcohol to the nearest county extension agent for confirmation if you're not sure about tracheal mites.

How to control

Here are some important methods to help you.

Use of menthol

Menthol, which you will get easily from most bee supply providers, is one tool for managing tracheal mites.

PESTS & BEE HIVE HEALTH

For the menthol to work, the temperature must be above 60° F. The bees inhale the vapor, which is thought to dehydrate the mites. To prevent contaminating honey, menthol must be eliminated during a nectar flood.

Oil extender patty

An oil extender patty is an acid solution for tracheal mites. There are two parts sugar and one part vegetable fat in this recipe.

•To use this, you will make a small four-inch patty and sandwich it between wax paper.

•Now you will cut along the edges of the wax paper to allow the bee's access to the patty.

•Place the oil patty in the middle of the hive shell, on top of the frames.

The sugar patty will attract the bees, who will then get oil on their skins. The oil serves as a chemical cloak, making it impossible for the tracheal mites to locate suitable bee hosts. Since the oil does not contaminate honey stocks, the oil patties are suitable for long-term use.

American Foulbrood (AFB)

The spore-forming bacterium Paenibacillus larvae cause American foulbrood (AFB), a very deadly bacterial disease of honey bee brood.

PESTS & BEE HIVE HEALTH

It is not a kind of stress-related illness, and it can infect any colony in a beehive, from the strongest to the weakest. The pre-pupal stage is when the infected brood normally dies. Infections that affect the majority of the brood will upset the colony and ultimately destroy it.

While AFB is not extremely infectious, bacterial spores can easily spread between hives and apiaries through common beekeeping practices like a kit exchange and the transfer of infected combs.

Signs

Some important signs that show that your hive is affected by them are:

•Bad odor in the brood, abnormally marked larvae

•Brood pattern that is irregular and patchy.

•The capping appears to be sunken and greasy.

How to control?

If you suspect AFB, you need to call your nearest Bee Inspector right away. Sadly, colonies, where AFB has been verified will be abandoned and burned.

European Foulbrood (EFB)

The European foulbrood (EFB) is a kind of bacterial disease.

PESTS & BEE HIVE HEALTH

It's classified as a stress disorder, and it's most common in the spring and early summer. It's not as bad as AFB, and colonies will recover from this infection. While EFB does not make spores, it often overwinters on combs. It enters the larva by infected brood food and multiplies rapidly in the larva's gut.

Signs

•Dead brood is typically younger and is dull white or light brown to almost black in color.

•The main consistency of the remains is rubbery and granular rather than elastic.

How to control?

Proper Antibiotics may be used to cure this honey bee disease.

Small hive beetle

The small hive beetle disease, which originated in Sub-Saharan Africa, is an invasive pest of beehives. In their native range, these beetles invade all honey bee colonies, but they do little harm and are rarely considered dangerous hive pests.

These beetles are generally treated as a secondary pest in the United States, inflicting excessive harm only if bee colonies have already been disrupted by other causes.

PESTS & BEE HIVE HEALTH

Signs

•Larvae burrowing into brood combs, eating brood and stocks

•Larvae clumping together in corners of frames or combs cells

•Small black beetles racing through the comb or hidden in small dark crevices of the hive

•Colonies of tiny 'rice grain' eggs in the hive's cracks and crevices

•Combs that are slimy or odor like rotting oranges

How to control?

It cannot be eradicated once it has established itself; beekeepers would need to introduce protection measures to deter invasions. It would be necessary to use an automated pest management method to handle this bee pest problem since there is no easy way to do so. The life cycle of the beetle must be disrupted or prevented in order to contain infestations.

Sacbrood Virus

Only honey bee larvae are affected by the Sacbrood Virus, which is slightly infectious inside the hive. The disease will usually be avoided by requeening in a colony.

PESTS & BEE HIVE HEALTH

After the break in the brood cycle, worker bees will destroy all infected larvae when no other larvae are present. Sacbrood has no known medical treatment.

Chalkbrood

Chalkbrood is a kind of fungal infection that enters brood through their food. It develops in the larva after the brood is locked in its nest. Chalkbrood larvae take on an opaque, white appearance.

Chalkbrood is typically located on the frames' fringes in late spring. If you see these tiny chalk nub bodies outside the hive entrance, where worker bees have dumped them, you know you have it.

There isn't a cure, but requeening can be one solution. The colony can survive if the workers can destroy any of the infected brood.

Paralysis

There are two kinds of paralysis usually found: chronic and acute. Eating pollen from some plants, as well as fermented stored pollen, may cause it. The bees lose their feathers, get greasy, and tremble as a result of this paralysis.

A colony can easily recover from paralysis on its own, but if it doesn't, you can requeen with a new strain since paralysis is also genetically transmitted.

PESTS & BEE HIVE HEALTH

Predators

Here are different types of predators.

Mice

Mice sometimes reach the hive during the winter when the bees meet or get into stored combs and destroy them by chewing the frames and combs to build their nest.

Skunks

Skunks eat a lot of bees at the entrance of the hive, usually at night. They can be controlled by the use of fences, cages, and poison.

Bears

The honeybees and brood in the hive are eaten by bears, which normally kill the hive. Electric fencing and traps are used to defend bee colonies in several bear countries.

Blood bath: When bees become their own enemy

In some cases, bees can also become their own worst enemy. They would be fighting for honey if it is exposed to them and the weather is cool.

PESTS & BEE HIVE HEALTH

This stealing, or battling, can become intense at times and spread from hive to hive. When all the bees in a colony are slaughtered, the honey is taken and brought to other hives easily.

This intensifies the robbery to the point that a cluster that was bringing honey into its hive is targeted, all of its inhabitants are killed, the honey is robbed once more, and the procedure is repeated. Only darkness or bad weather can end robbery.

Colony Collapse Disorder (CCD)

An unknown threat has been affecting bee colonies in many parts of the world since 2006. Beekeepers began to experience massive losses in their colonies. Up to 90 percent or more of the hives were lost. The extent of the losses was especially very confusing. There were no signs of the traditional risks in at least half of these situations. There were no apparent symptoms of illness, elevated mite counts, or any other obvious cause.

Colony Collapse Disorder is the name assigned to this series of events. The bee population is decreasing due to CCD and other challenges. Scientists are still looking for the root of the problem and what potential solutions can be. Many people claim that industrial pesticides, especially neonicotinoids, are the main reason behind this problem.

Breeding bees is one of the most popular solutions. So there really is a need for you and many more to start a beekeeping, with this current bee crisis.

CHAPTER 5: Making Massive Money with Honey Bee Products

MONEY WITH HONEY BEE PRODUCTS

Pollen, propolis, royal jelly, queens, bees, venom, and bee larvae are some important bee products. But honey and wax are the most famous primary bee products. The majority of these products can be used immediately, as they are produced by bees. However, there are other uses where bee products can be used as a component of another product.

The addition of bee products to other products typically increases the relative importance of consistency of these secondary products due to their quality and often almost magical prestige and features. Many beekeeping operations will benefit more as a result of this.

The following are the most popular bee products used by humans:

Honey

Honey bees are famous for collecting nectar from flowers and store it in a honey crop, which looks like a gut. When the bee returns to the hive, another bee collects the nectar and spreads it over the wax honeycomb to aid in the evaporation of its moisture. The second bee also adds invertase, an enzyme that assists in the breakdown of sugar molecules. It is enclosed in a cell with a wax cap until it has thickened.

MONEY WITH HONEY BEE PRODUCTS

Pollen

Bees often gets covered in pollen as they extract nectar from plants, which they then supplement with several ferments, hormones, and antibiotics before depositing it in honeycomb cells. We get extracted honeycomb pollen or bee bread when we remove these pollen balls from a honeycomb cell. We can also get fresh pollen if we put a pollen trap at the hive's entrance. Pollen loads spill off bees' legs as they attempt to squeeze through the net of a pollen trap.

Pollen has a high protein. It has all of the basic amino acids, as well as a variety of fatty acids, vitamins B, C, D, E, and K, as well as provitamin A. It is advised that pollen be soaked before eating it.

Propolis

Propolis, in particular, is a unique treasure of the beehive as it is a kind of natural antibiotic. When feeding larvae, bees collect resin from a number of trees and shrubs and mix it with pollen pellets. It has contained over 360 compounds. Antibacterial and antifungal effects are found in propolis.

Royal jelly

The larvae are fed by royal jelly, which is rich in protein. The queen larva gets more jelly than others, allowing her to grow larger than the other bees.

MONEY WITH HONEY BEE PRODUCTS

It contains carbohydrates, fats, amino acids, vitamins, minerals, and proteins and is made from digested pollen and honey.

Beeswax

Wax is usually developed by worker bee glands and is used to create honeycomb and seal the tops of honey-filled cells.

Fatty-acid are the key ingredient of beeswax, which comprises over 300 natural compounds. Wax is almost pure white when it is raw, but it gradually becomes a yellowish-brown hue. It has a good sugar, propolis, and pollen-like smell. Wax is widely used in creams because it softens the skin and has antibacterial effects.

The chewing of capping, i.e., the wax covering over honey, is well-known, and beeswax-based thermal therapies, which are typically used during a massage or physiotherapy, are now becoming more common.

Bee Venom

A unique combination of proteins makes up the venom which is found in bee stings. Humans can get benefit from the venom, according to new studies. Bee venom is also used in medicine to pacify people who are allergic to bee venom. Bee venom is also used to cure a variety of problems and ailments all around the world, but only in strict medical observation.

MONEY WITH HONEY BEE PRODUCTS

Bee venom has recently gained popularity in the cosmetics industry. Since it's meant to be a natural Botox supplement, it's used in different creams and serums.

Other products and services

As we know that honey is the most common source of income for beekeepers, there are also a variety of other ways to get profit from bees. Learning each of these skills takes time, and not everyone is prepared for any of them, but you can have an idea about other uses of products and what you want to do with your bee colonies.

Here are some of the most successful ways to make money from your beekeeping business:

Queen cells, virgins, and queens

Northern beekeepers like to have queens that are free of African genes. Texas beekeepers also want the same queens for themselves. Only very few African beekeepers require African queens that have been picked for less swarming and higher honey production. You can also raise queens for yourself and sell them to other beekeepers in your area to get a high profit.

MONEY WITH HONEY BEE PRODUCTS

Deliver Pollination services

Many farmers pay to have new hives or temporarily move some hives to their fields to provide pollination services. The hives are typically employed for a three to five-week duration to complete their task. The majority of farms in need of pollination facilities are in California, Texas, and Florida.

Almonds, sunflowers, and canola are the crops with the greatest demand for pollination services. In 2012, the pollination industry was worth $655 million, according to the USDA.

Provide hive kits for beginners

Some beekeepers can also earn their profits by supplying beginner kits to new beekeepers. You might also try selling an all-in-one starter kit that includes the hive, bees, and some other essentials like the smoker and important safety equipment. You are not limited to local consumers, and bees can be sent by postal systems in special screened shipping containers.

CHAPTER 6:
How to Harvest
Bee Products

HOW TO HARVEST BEE PRODUCTS

Is there anything more enjoyable than your own honey? Here are few suggestions to think about.

What is the right time to harvest your honey?

The most suitable time for the extraction of honey is during the summer, but if your hive is ready, you can harvest in spring. When all of your bees' cells have been capped, you'll know it's time to harvest the honey.

Know the difference between Brood cells and honey cells

Worker bees usually cap the cells to create food for the winter period. Capped cells, on the other hand, can produce larvae, so understanding the difference between brood cells and honey cells is very important before any extraction.

Honey Supers

Capped honey cells are most often used in supers, which are kind of boxes at the top of the hive. You will find them separated from the rest of the boxes by a queen excluder. The excluder helps to stops the queen from laying eggs in the top supers.

At least 90% of the cells must be capped

The best way to find out that you're not extracting your honey too early or too late is to look for capped cells. You have to make sure that at least 90% of the cells have been capped off by nurse bees.

HOW TO HARVEST BEE PRODUCTS

Trust the bees' instincts; they know when the moisture level is just right for sealing the cells.

What if you extract before the right time?

If you start extracting honey before the right time, it will have moisture and will ferment in storage.
Keep in mind that your bees must have sufficient time to search in order to fill their hive. Observing the atmosphere will assist you in determining honey consumption, which is a period of time during which bees have access to all they need in order to produce honey.

What if you harvest honey too late?

It's important to understand that your bees are busy for a reason. The honey they're making in their hive is supposed to last for them all winter. As a result, having an abundance of honey is a very good thing in the winter. On the other hand, Bees do not avoid filling their cells with honey until the foraging season is over.

Swarming may occur

In the spring, if there is honey left in any hive, there will not be enough space for bees to store it; as a result, they may run out of space, and a swarm will occur.

HOW TO HARVEST BEE PRODUCTS

Difficult to extract honey

Furthermore, if you haven't performed your extraction correctly and are too late to extract, honey can be hardened in the cells, making extraction more difficult.

How to harvest your honey?

Time to start harvesting your honey once you've successfully maintained your honey bee colony. Extraction is the term used for honey harvesting, which occurs once a year on average. Let's start with the tools:

Equipment needed in the extraction process

During the extraction process, beekeepers use a variety of helpful methods to make this process a bit simpler.

•Knife for uncapping

•Comb for uncapping

•Container for collecting wax

•Extractor machine

•Double honey sieve made of stainless steel

•Containers for honey

HOW TO HARVEST BEE PRODUCTS

Now it's time to harvest honey!

But first!

•You need to wear protective clothing, gloves, shoes, and a hat before beginning the process.

•Do not try to speed up the honey harvesting process. The bees will not get disturbed if you make soft, calm motions rather than huge, exaggerated ones.

•Be sure you're not using any perfumes, colognes, aftershaves, or other scented items, as these can attract curious bees, making it more difficult.

Secondly,

Always keep in mind that you should not take all honey from the hive, as it is bee food. As a general rule, leave at least 85 pounds of honey for your bees and take the remaining.

Step 1: How to Get Bees Out of the Supers? There are a few choices for safely removing bees from their honey frames, so let's have a look at them:

Smoker and Bee Brush

An easy way of removing bees from the hive is a puff of smoke and shaking or gently scratching the frame with the bee brush.

HOW TO HARVEST BEE PRODUCTS

When you plan to brush the bees off their frames, be careful and always use an upward rather than downward stroke to avoid hurting or killing them.

Fume Boards

A fume board is the easiest way to extract honey from a beehive. A fume board resembles a typical top or outer hive cover on the exterior, but on the inside, absorbent material is sprayed with a non-toxic substance that normally bees do not like.

Place the fume board on top of the honey super that you want to harvest when it's finished. After a few minutes, the bees will flee due to the odor and vacate the honey frame, allowing you to empty the honey box without any disruption.

Step 2: Harvesting Honey

It is time to start harvesting now as you've separated your bees from the supers and have all of your equipment set.

Make sure your honey collection jar is in place under the extractor's spigot, and your wax collection container is ready to use.

Warm Your Knife

Now place a honey-filled frame on top of your wax collection jar, start sawing the caps open from top to bottom, and letting them fall into your container.

HOW TO HARVEST BEE PRODUCTS

Make sure your uncapping knife should be warm but not hot.

Using your honeycomb

You can also use the comb for extracting the honey and opening the wax cap by softly stabbing the stubborn cells.

Place your frame in the honey extractor.

Depending on the sort of extractor you're using, load your frames into it and manually spin it or activate the device electronically.

Honey will appear to come out of the extractor's walls and gather at the bottom. The honey would then flow slowly out of the spigot and into your honey collection tub.

If you're manually turning your extractor, you'll need to keep an eye on the amount of honey that's accumulating in the rim.

Drain and filter the honey from the extractor

You will find on the bottom of the extractor a drain with a shut-off valve. When extracting honey, the valve must be left open to allow the honey to drain. You need to place the stainless double honey sieve on top of a 5-gallon bucket to suck out wax particles and honeybee parts. To filter your honey, you need only these things.

HOW TO HARVEST BEE PRODUCTS

Let it rest

Once you've drained your frames and filled your honey bucket, cover it with a lid and set it aside for the next 24 hours to allow bubbles to rise before saving them in a bottle.

Once your honey has rested for 24 hours, quickly add it into jars and cover them with lids. Honey that has been exposed to the air for an extended period of time will collect moisture, which will lead to fermentation in the future.

How to extract your honey Beeswax?

Honey extraction is a very interesting process, but don't forget about the important wax you scraped into your straining container!

Wax is used in a variety of products, including salves, candles, sweets, cosmetics, and more. If you aren't crafty, someone else may be interested in buying your wax, so harvesting will help you and bring you profit.

Procedure to follow

You need to follow these steps in order to get a wax.

Step 1: After uncapping the cells, strain the honey carefully to remove any wax that has remained.

HOW TO HARVEST BEE PRODUCTS

Step 2: To dissolve the remaining honey, cover the remaining wax with warm water and swirl it around; please don't overheat the wax with the water.

Step 3: Drain the water from the beeswax and put it in a double boiler; you need to add much as you would use for candy making.

Step 4: Melt the wax in a low-heat setting.

Step5: Now strain the wax through a fine strainer to eliminate any unwanted elements left behind by the bees. This method can be repeated many times to guarantee that the wax is pure.

Step 6: Pour the warm beeswax into a jar so it can easily be removed when needed. Keep refrigerated until ready to use.

How to collect Royal jelly?

Normally beekeepers only collect honey, but they can also collect royal jelly sometimes. Worker bees are responsible for feeding royal jelly to the queen bee to help her increase in size. It's also a well-known consumer food with several health benefits. The following steps are necessary for harvesting royal jelly:

•Removing the cells

•Extracting the jelly

•Correctly storing it

HOW TO HARVEST BEE PRODUCTS

Step 1: Getting out the Cells from the hive
First of all, wear your protective clothes to start the process.

The Ideal time to harvest royal jelly

The perfect time for royal jelly extraction is when larvae are three days old. This will produce the royal jelly in large quantities.

Pull the frames slowly and brush the Bees.

Now remove those frames that have royal jelly cells and begin brushing the bees away. To stop disturbing the bees, remain quiet and step slowly and softly. Be patient; brushing the bees away will take some time to get rid of all of them.

Step 2: Extracting Royal Jelly from the frames
For extraction, you need to use a sharp knife.

Use a sharp knife

Cut the narrow portion of each cell with a knife. Each of the cells has a narrow end that contains the larvae and jelly inside. Break these ends off with a sharp knife. When you slash, holding the knife's flat edge level with the cell's edge.

Pull the larvae out of the cells

You need to pull the larvae out of the cells with tiny forceps. In the middle of each cell, look for the slightly coiled larvae.

HOW TO HARVEST BEE PRODUCTS

It will resemble a fat, little worm. It's important to get rid of the larvae before harvesting the royal jelly.

Use spatula or pipette

Now remove the jelly with the help of a spatula.

Use a dark glass storage container to store the royal jelly

Place the jelly in a glass jar with a lid as you collect it. If you're gathering a huge quantity of jelly at once, you may want to place the bottle on ice to keep it fresh.

Point to note: If left unrefrigerated, royal jelly spoils in a couple of hours, so you may want to pick smaller quantities at a time and put them in the refrigerator. Always use a freezer to safe the vial and leave about 0.5-inch place for the jelly to expand. You can save it for almost 18 months in a freezer.

How to collect the Bee Bread or Bee Pollen?

Bee pollen is high in protein and serves as a building block for a bee colony. According to national statistics, Beehives can yield anywhere from 1 kilogram to 7 kilograms of bee pollen per year, averaging 50 to 250 grams per day.

HOW TO HARVEST BEE PRODUCTS

Pollen traps to the entrance point of the hive

Pollen traps are attached to the hive's entry points by beekeepers. The pollen trap softly scrapes any pollen from the bee's legs as they re-enter the hive with their set.

Collecting the pollen grains on a separate tray

The pollen is then gathered on a different tray or catchment underneath the hive for a quick harvest. Bee pollen harvesting in this way mobilizes bees and increases the number of foragers. Pollen grains can be light yellow, brown, or even black in color when harvested.

Now you officially become a Beekeeper!

It's fair to assume that if you've harvested your first batch of honey, you finally become a beekeeper. You may find this process hard to start, but it gets easier after a few years.

CHAPTER 7: Beekeeping Safety

BEEKEEPING SAFETY

So if you're new in the beekeeping business, you definitely have a lot of goals in mind, such as understanding the differences between queens and worker bees, knowing about eggs and brood, discussing hive positions, and learning everything that you can about your bees.

Another important target that you should keep in mind is your own protection during bee inspection. Although beekeeping is quite an interesting venture, it is important to note that it is not without risk.

You'll be living in close quarters with tens of thousands of stinging insects that hate disruptions and will valiantly protect their home if they consider you as a threat. It's a good idea to do your research about safety tricks and treat your passion with care. Here are a few tips to keep you safe when working in the apiary.

Do your work carefully

Bees do not like surprises, such as disturbance in their happy home. So, you will continue to make the routine "checkup" or make it easy for the bees by working carefully. Have you ever seen how tapping on a hive produces a brief burst of rapid buzzing? Since bees hate noisy sounds, always try to keep the sound as low as possible, with little bumps and bangs.

BEEKEEPING SAFETY

You'll have a better time with fewer bees taking flight around you if you can give the "risk" warning (your smoker plays a key role in this). Take your time, but don't leave your hives open for a longer period of time. Work consciously, keeping in mind all the risks.

Inspect During pleasant Weather

Always select a warm, sunny afternoon if you're going to do simple routine work in your hives, such as tracking eggs and brood or testing the honey and pollen. The main reason behind this is, on a sunny day, your hive will have fewer bees inside, so more of the workers will be out "working."

So it will be easier and safer to work in this less-crowded hive because you'll be less likely to disturb the bees. Furthermore, some beekeepers claim that a colony is usually "angrier" on rainy or stormy days, so they don't work in the hive on a rainy day.

Wear proper protective cloths

Many beekeepers like to work without hats, in normal clothing, or even without a veil. Does it mean that they're not scared of being stung? Perhaps not at this stage. They've probably dealt with bees for so long that they can easily deal with them, and many experienced beekeepers appreciate the independence that comes with skipping gloves and other protective gear.

BEEKEEPING SAFETY

But this is not right. It's better to wear all protective clothing if you are a novice. Protective garments will give you peace of mind, helping you to focus more deeply on your job and learning about your bees' habits. It should also give you more confidence in going to work, particularly when bees start buzzing around your head and crawling up your hands.

Keep your Hive Tidy

It is also important to keep a clean environment around your hives. It will keep both you and your bees healthy, which can surprise you. Always empty your hive boxes, or make sure that old doors and comb should not be left standing, as this honey-scented stuff will lure skunks, raccoons, and even bears. These creatures are a danger to not only your hives but to you as well. Therefore it is very important to keep things in proper order.

Maintain the cleanliness of your equipment

Your bees' wellbeing can be influenced by how you treat your hives and other supplies. You need to be very careful while using your equipment. A good cleaning can play an important role in avoiding the transmission of infection.

Be mindful that thorough cleaning and sterilizing of the hives and machinery can be very difficult. Before you begin, double-check that you have all of the appropriate supplies and equipment for the job. Have you ever wondered when do we disinfect our machines, supplies, and boxes?

BEEKEEPING SAFETY

•When you carry your boxes for storage at the end of the season

•If the colonies are infected or diseased

•Every time you think about re-using one of your tools

•When you decide to transfer things from one colony to another.

•If the tools get old or infected with bacteria.

Lyme disease

Despite its many benefits, beekeeping comes with a number of possible threats and dangers. More recently, the high risk of contracting diseases like Lyme has become a major danger for beekeepers, especially in the northeastern United States.

Ticks can spread the bacteria that cause Lyme disease. As a result, beekeepers are strongly advised to search themselves for ticks when returning from the bee yard.

•Tick tests should be complete and conducted in a special setting.

•Tick-removal products are also available in the market to aid with tick removal.

•Wearing light-colored long-sleeved tops and long trousers to make ticks easier to find.

BEEKEEPING SAFETY

Finally!

Without any doubt, beekeeping is all about learning more for the sake of your bees, the environment, and yourself. Experience, honey, and many other advantages that even you can imagine would be your reward. You will continue to enjoy your time in the hive as long as you are mindful of the issue that may arise and can take the necessary precautions to remain healthy.

Follow all of these procedures and you can take the sting out of working with bees.

Chapter 8: Starting a

Beekeeper Business

Step by Step

GETTING STARTED IN BUSINESS

Starting a Beekeeping business can seem like a big challenge, but with a step by step blue print to help you get started with the basics things will be a snap.

To begin with you might want to visit other beekeeper's to get a feel for what you like and may not like about the business. Study what is already working in this marketplace and duplicate their success.

There are over thirty million home-based businesses in the United States alone.

Many people dream of the independence and financial reward of having a home business. Unfortunately they let analysis paralysis stop them from taking action. This chapter is designed to give you a road map to get started. The most difficult step in any journey is the first step.

Anthony Robbins created a program called Personal Power. I studied the program a long time ago, and today I would summarize it, by saying you must figure out a way to motivate yourself to take massive action without fear of failure.

2 Timothy 1:7 King James Version

"For God hath not given us the spirit of fear; but of power, and of love, and of a sound mind."

GETTING STARTED IN BUSINESS

STEP #1 MAKE AN OFFICE IN YOUR HOUSE

If you are serious about making money, then redo the man cave or the woman's cave and make a place for you to do business, uninterupted.

STEP #2 BUDGET OUT TIME FOR YOUR BUSINESS

If you already have a job, or if you have children, then they can take up a great deal of your time. Not to mention well meaning friends who use the phone to become time theives. Budget time for your business and stick to it.

STEP #3 DECIDE ON THE TYPE OF BUSINESS

You don't have to be rigid, but begin with the end in mine. You can become more flexible as you gain experience.

GETTING STARTED IN BUSINESS

STEP #4 LEGAL FORM FOR YOUR BUSINESS

The three basic legal forms are sole proprietorship, partnership, and corporation. Each one has it's advantages. Go to www.Sba.gov and learn about each and make a decision.

Sole proprietorship: A sole proprietorship, also known as a sole trader, is owned by one person and operates for the profit of that person. The owner runs the establishment alone or may hire employees. A sole proprietor has unlimited liability for all obligations incurred by the business, whether from operating costs or judgments against the business. Every assets of the business belong to a sole proprietor, including, for example, a desk, laptop, all inventory, business equipment, or retail equipment, as well as any real estate owned by the sole proprietor.

Partnership: This type of business owned by 2 or more people. In the majority of partnerships, each partner has unlimited liability for the debts incurred by the business. The 3 most common types of for-profit partnerships are general partnerships, limited partnerships, and limited liability partnerships.

GETTING STARTED IN BUSINESS

Corporation: The owners of a corporation have limited liability and the business has a separate legal personality from its owners. Corporations can be either government-owned or privately owned, and they can organize either for profit or as nonprofit organizations.

A privately owned, for-profit corporation is owned by its shareholders, who elect a board of directors to direct the corporation and hire its managerial staff. A privately owned, for-profit corporation can be either privately held by a small group of individuals, or publicly held, with publicly traded shares listed on a stock exchange.

STEP #5 PICK A BUSINESS NAME AND REGISTER IT

One of the safest ways to pick a business name is to use your own name. By using your own name you don't have to worry about copy right violations.

However, always check with an Attorney or the proper legal authority when dealing with legal matters.

GETTING STARTED IN BUSINESS

STEP #6 WRITE A BUSINESS PLAN

This would seem like a no brainer. No matter what you are trying to accomplish you should have a blueprint. You should have a business plan. In the NFL about seven headcoaches get fired every season. So in a very competetive business, a man with no head coaching experience got hired by the NFL's Philadelphia Eagles. His name was Andy Reid. Andy Reid would later become the most successful coach in the team's history. One of the reasons the owner hired him, was because he had a business plan the size of a telephone book. Your business plan does not need to be nearly that big, but if you plan for as much as possible, you are less likely to get rattled when things don't go as planned.

STEP #7 PROPER LICENSES & PERMITS

Go to city hall and find out what you need to do, to start a home business.

STEP #8 PUT UP A WEB SITE, SELECT BUSINESS CARDS, STATIONERY, BROCHURES

This is one of the least expensive ways to not only start your business but to promote and network your business.

STEP #9 OPEN A BUSINESS CHECKING ACCOUNT

Having a separate business account makes it much easier to keep track of profit and expenses. This will come in handy, whether you decide to do your own taxes or hire out an professional.

STEP #10 TAKE SOME SORT OF ACTION TODAY!

This is not meant to be a comprehensive plan to start a business. It is meant to point you in the right direction to get started. You can go to the Small Business Administration for many free resources for starting your business. They even have a program(SCORE) that will give you access to many retired professionals who will advise you for free! Their web site: **www.score.org**

Chapter 9:
How to Write a
Business Plan

HOW TO WRITE A BUSINESS PLAN

Writing a business plan is one of the most important steps you can take to ensure that you Beekeeper's business will succeed.

"Study to shew thyself approved unto God, a workman that needeth not to be ashamed, rightly dividing the word of truth."

2 Timoth 2:15 King James Version

To make your business plan a successful one, study other successful beekeeping businesses. Do your research.

* What are their best selling products?

* How much does it cost to produce those products?

* What is the annual revenue?

* How to they promote their business?

More and more beekeeper businesses are using YouTube, the free video sharing platform, to promote their business.

Look at the amount of subscribers the channel has and the amount views on their videos to determine who are the best beekeeping YouTubers.

Then view their YouTube channel and business website to help give you ideas for what your business will need in it's plan for success.

HOW TO WRITE A BUSINESS PLAN

Millions of people want to know what is the secret to making money. Most have come to the conclusion that it is to start a business. So how do you start a business? The first thing you do to start a business is to create a business plan.

A business plan is a formal statement of a set of business goals, the reasons they are believed attainable, and the plan for reaching those goals. It may also contain background information about the organization or team attempting to reach those goals.

A professional business plan consists of eight parts.

1. Executive Summary

The executive summary is a very important part of your business plan. Many consider it the most important part because this part of your plan gives a summary of the current state of your business, where you want to take it and why the business plan you have made will be a success. When requesting funds to start your business, the executive summary is a chance to get the attention of a possible investor.

2. Company Description

The company description part of your business plan gives a high level review of the different aspects of your business. This is like putting your elevator pitch into a brief summary that can help readers and possible investors quickly grasp the goal of your business and what will make it stand out, or what unique need it will fill.

3. Market Analysis

The market analysis part of your business plan should go into detail about your industries market and monetary potential. You should demonstrate detailed research with logical strategies for market penetration. Will you use low prices or high quality to penetrate the market?

4. Organization and Management

The Organization and Management section follows the Market Analysis. This part of the business plan will have your companies organizational structure, the type of business structure incorporation, the ownership, management team and the qualifications of everyone holding these positions including the board of directors if necessary.

5. Service or Product Line

The Service or Product Line part of your business plan gives you a chance to describe your service or product. Focus on the benefits to the customers more than what the product or service does. For example, a air conditioner makes cold air. The benefit of the product is it cools down and makes customers more comfortable whether they are driving in bumper to bumper traffic or are sick and sitting in a nursing home. Air Conditioners fill a need that could mean the difference between life and death. Use this section to state what are the most important benefits of your product or service and what need it fills.

6. Marketing and Sales

Having a proven marketing plan is a essential element to the success of any business. Today online sales are dominating the marketplace. Present a strong internet marketing plan as well as social media plan. YouTube videos, Facebook Ads and Press Releases all can be part of your internet marketing plan. Passing out flyers and business cards are still an effective way to reach potential customers.

Use this part of your business plan to state your projected sales and how you came to that number. Do your research on similar companies for possible statistics on sales numbers.

HOW TO WRITE A BUSINESS PLAN

7. Funding Request

When you write the Funding Request section of your business plan, be sure to be detailed and have documentation of the cost of supplies, building space, transportation, overhead and promotion of your business.

8. Financial Projections

The following is a list of the important financial statements to include in your business plan packet.

Historical Financial Data

Your historical financial data would be bank statements, balance sheets and possible collateral for your loan.

Prospective Financial Data

The prospective financial data section of your business plan should show your potential growth within your industry, projecting out for at least the next five years.

You can have monthly or quarterly projections for the first year. Then project from year to year.

Include a ratio and trend analysis for all of your financial statements. Use colorful graphs to explain positive trends, as part of the financial projections section of your business plan.

HOW TO WRITE A BUSINESS PLAN

Appendix

The appendix should not be part of the main body of your business plan. It should only be provided on a need to know basis. Your business plan may be seen by a lot of people and you don't want certain information available to everybody. Lenders may need such information so you should have an appendix ready just in case.

The appendix would include:

Credit history (personal & business)

Resumes of key managers

Product pictures

Letters of reference

Details of market studies

Relevant magazine articles or book references

Licenses, permits or patents

Legal documents

Copies of leases

HOW TO WRITE A BUSINESS PLAN

Building permits

Contracts

List of business consultants, including attorney and accountant

Keep a record of who you allow to see your business plan.

Include a Private Placement Disclaimer. A Private Placement Disclaimer is a private placement memorandum (PPM) is a document focused mainly on the possible downsides of an investment.

Chapter 10:

$5 Million Dollars to Fund Your Business

$5 MILLION DOLLARS

When I was first starting out in business, a mentor told me to stick with the United States government when it came to getting money.

Did you know that the United States government gave billionaire, yes billionaire Elon Musk over 5 billion dollars in government grants? If a person who is already a billionaire and likely pays zero in taxes, qualifies to get your tax dollars you should at least look into getting some of your money yourself.

Since that time, I have gotten thousands of dollars in government grants and loans.

When it comes to finding money from the government for your beekeeping business, or any business for that matter, Matthew Lesko is the best there is.

This is Matthew Lesko's website:

https://www.free.lesko.com/leskohelp24341993

As of this writing, it has a free download for 32 government grant applications. This could change, but regardless there is always quality government money information on this website that is not common knowledge.

Below is the website for the Catalog of Federal Domestic Assistance. It has a full listing of all the government assistance programs available.

https://recovery.fema.gov/glossary/CFDA

$5 MILLION DOLLARS

Loans guaranteed by the Small Business Administration can be as little as $500 to as big as $5 Million Dollars!

The money can be used for a variety of business needs, including the purchase of long-term fixed assets and for operating expenses. Some loan programs do have restrictions on how the loan money can be used, so you will have to check with a Small Business Administration approved lender when looking for a loan. The lender can match you with the correct loan for your business needs.

Working Capital

Working capital could be seasonal financing, export loans, revolving credit, and refinanced business debt.

Fixed Assets

Fixed Assets could be office equipment, property, tools, machinery, business equipment, construction, and remodeling.

$5 MILLION DOLLARS

Eligibility requirements

Lenders and loan programs have distinctive eligibility guide lines. Basically, eligibility is related to what a business does to receive its funding, the character of its ownership, and location of the businesses operation. Usually, businesses must meet size standards.

What is a small business size standard?

A business size standard, under most circumstances is stated in the number of employees or average yearly receipts, and represents the biggest size that a business (including its subsidiaries and affiliates) may be to remain classified as a small business for Small Business Administration and government contracting programs. The definition of "small" can be different in different industries.

How to calculate your small business size

Size standards are mostly based on the average annual receipts or the average number of employees.

$5 MILLION DOLLARS

Eligibility requirements

You must be able to repay the loan. You must have a credible business objective. Individuals with bad credit may still qualify for business startup money. Lenders will give you a list of the lending guide lines and requirements for your loan. Here are a few more.

Be a for-profit business

The business is properly registered and performs as a legal business.

Do business in the U.S.

The business is physically located and operates in the United States and or its territories.

You Have invested equity

You the business owner has invested your own time or finances into the business.

$5 MILLION DOLLARS

Eligibility requirements

Exhaust financing options

The business cannot get money from any other financial lender.

Loans for exporters

Most United States banks view loans for exporters as risky. This can make it more difficult for you to get loans for things like day-to-day operations, advance orders with suppliers, and debt refinancing. That's why the Small Business Administration came up with programs to make it easier for United States small businesses to get loans for an export business.

To learn how the SBA can help you get an export loan, contact your local Small Business Administration International Trade Finance Specialist or the Small Business Administration's Office of International Trade.

https://www.sba.gov/funding-programs/loans

Chapter 11:

Colossal Cash

Crowdfunding

COLOSSAL CASH CROWDFUNDING

" yet ye have not, because ye ask not. Ye ask, and receive not, because ye ask amiss"
James 4:2-3 King James Version

Sometimes the easiest way to get something is to ask, and ask politely. Crowdfunding is another way you could use to fund your beekeeping business simply by asking people to give you money.

If you have never heard of Crowdfundin it is "the practice of funding a project or venture by raising many small amounts of money from a large number of people, typically via the internet."

In this chapter you are going to get a overview of this process so that you can add this form of funding to your business arsenal.

Here are some of the top Crowdfunding websites when it comes to Startups:

* https://www.crowdsupply.com/

* https://experiment.com/

* https://chuffed.org/us

There are plenty of other websites that I will go into more detail later on in this chapter but if the idea of getting free money gets you excited, these will get your started!

COLOSSAL CASH CROWDFUNDING

In 2015 over $34 billion dollars was raised by crowdfunding. Crowdfunding and Crowdsourcing roots began in 2005 and they help to finance or fund projects by raising money from a large number of people, usually by using the internet.

This type of fundraising or venture capital usually has 3 components. The individual or organization with a project that needs funding, groups of people who donate to the project, and a organization sets up a structure or rules to put the two together.

These websites do charge fees. The standard fee for success is about %5. If your goal is not met, there is also a fee.

Below is a list of the top Crowdfunding websites according to myself and Entrepreneur Magazine Contributor Sally Outlaw.

COLOSSAL CASH CROWDFUNDING

https://www.indiegogo.com/

Started as a platform for getting movies made, now helps to raise funds any cause.

http://rockethub.com/

Started as a platform for the arts, now it helps to raise funds for business, science, social projects and education.

http://peerbackers.com/

Peerbackers focuses on raising funds for business, entrepreneurs and innovators.

https://www.kickstarter.com/

The most popular and well known of all the crowdfunding websites. Kickstarter focuses on film, music, technology, gaming, design and the creative arts. Kickstarter only accepts projects from the United States, Canada and the United Kingdom.

COLOSSAL CASH CROWDFUNDING

http://group.growvc.com/

This website is for business and technology innovation.

https://microventures.com/

Get access to angel investors. This website is for business startups.

https://angel.co/

Another website for business startups.

https://circleup.com/

Circle up is for innovative consumer companies.

https://www.patreon.com/

If you start a YouTube Channel (highly recommended) you will hear about this website frequently. This website if for creative content people.

COLOSSAL CASH CROWDFUNDING

https://www.crowdrise.com/

"Raise money for any cause that inspires you."
Landing page slogan speaks for itself. #1 fundraising
website for personal causes.

https://www.gofundme.com/

This fundraising website allows for business, charity,
educatiion, emergencies, sports, medical, memorials,
animals, faith, family, newlyweds etc...

https://www.youcaring.com/

The leader in free fundraising. Over $400 million
raised.

https://fundrazr.com/

"FundRazr is laser-focused on eliminating the
guesswork of raising money online for your
campaign. They use technology and social media
guidance make telling your powerful story easy;
sharing it with the widest community simple; and
collecting the money worry-free. "

Chapter 12: Advertise to a Billion People for Free!

YouTube Video Marketing Overview

YouTube is an amazing platform to promote your Beekeeping business. However it is critical that you create video content of value. Content that entertains and informs the viewer. Some of the best content on YouTube goes unviewed, because YouTube success depends on you knowing the platforms success secrets.

After you read this chapter you will have a good overall understanding of YouTube, but the competition has only grown and I suggest you view, and maybe even subscribe to these YouTube channels listed below, to keep up to date on the latest changes and strategies for having success with a YouTube channel that can help to promote your beekeeping business.

Brian Dean
https://urlzs.com/ex5HN

Nick Nimmin
https://urlzs.com/oaGnq

Both of these guys have a ton of up to date content for growing your YouTube channel, and this will help you beekeeping business to grow and sustain that growth!

YouTube Video Marketing Overview

Million Dollar Video Marketing

When you read the title of this chapter you may have thought the term "Million Dollar" was hyperbole. However the beauty of video marketing is that it can be done for free, and that there really are several people who make millions of dollars just on their YouTube video's alone. Meaning that they allow ads to be placed on them and they get paid a portion of what google gets from businesses that runs the ads.

Since they are only getting a portion of what is being paid, that means if they make a million dollars, the video's actually produced multi-millions of dollars in ad revenue.

Here are a list of YouTube Millionaires as reported by Forbes magazine in the 20 December 2016 issue.

Youtube name/channel	2016 Income
1. Pewdiepie	$15 Million

Makes video's of himself playing video games and making crude comments on girls dancing.

2. Atwood	$8 Million

YouTube Video Marketing Overview

Promotes products and tours with other Youtubers.

   3. Lilly Singh $7.5 Million

Makes comedy skits mostly featuring herself talking about her parents and relationship issues.

YouTube name/channel	2016 Income
4. Smosh	$7 Million
Comedy Duo.	
5. Rosanna Pasino Nerdie Nummies	$6 Million
Baking show	
6. Markipler	$5.5 Million
Comments on Video Games.	
7. German Garmendia	$5.5 Million
Got a publishing deal from his YouTube channel	
8. Miranda Sings	$5 Million
Comedian	

YouTube Video Marketing Overview

9. Collen Ballinger $5 Million

Comedian

10. Tyler Oakley $5 Million

Makes a diary. LGBT Activist

And these are just some the the top earners. There are many more making $50,000 a month talking about movies, how to put on make up or video taping a day at an amusement park.

A Few Keys to Video Marketing Success

1. Commitment

While many of the top YouTubers are funny, they take their business seriously. One of the first things you have to understand is that there is commitment needed to be successful on YouTube.

Many of the successful YouTubers put up video's daily! One such YouTuber is Grace Randolph (Beyond the Trailer). Grace comments on movie news and movie trailers. She typically uploads 1-3 video's a day.

YouTube Video Marketing Overview

2. Research

Just putting up a video will not guarantee views. You have to put in research for every video. Research if the topic is popular or trending. Research what keywords you should use in your video. Research the success of other video's. Skip the research, skip the success.

3. Popularity

There are certain topics on YouTube that are extremely popular. Star Wars, Disney, Scantily clad women, video games, comedy. Know the level of your topics popularity and try to use keyword planning to max out the highest possible level. Some educational material is extremely valuable, but not popular.

ZERO COST MARKETING OVERVIEW

This is a zero cost online marketing plan for any business, cause or idea you wish to promote. This plan will show you step by step how to use online marketing featuring YouTube and Article Marketing to get free advertising for any product. In addition, this chapter will show you how to use this zero cost marketing plan to create a passive income stream.

YouTube Video Marketing Overview

A Few Key Definitions

YouTube is a video-sharing website headquartered in San Bruno, California, United States. The service was created by three former PayPal employee in February 2005. In November 2006, it was bought by Google for 1.65 Billion dollars. According to the Huffington Post, YouTube has 1 billion active users each month. Or nearly one out of every two people on the internet.

AdSense (Google AdSense) is an advertising placement service by Google. The program is designed for website publishers who want to display targeted text, video or image advertisement on website pages and earn money when the site visitors view or click the ads.

Hyperlink is a link from a hypertext file or document to another location or file, typically activated by clicking on a highlighted word or image on the screen.

Black Hat

In search engine optimization (SEO) terminology, black hat SEO refers to the use of aggressive SEO strategies, techniques and tactics that focus only on search engines and not a human audience, and usually does not obey search engines guidelines.

YouTube Video Marketing Overview

Getting Started

You get started by opening up a YouTube account. Go to www.YouTube.com and follow the step by step instructions. Then you open up a AdSense account. The AdSense account will take about a week to open. AdSense is linked to your YouTube account and land bank account. AdSense will use your 9 digit routing number to deposit a small amount of money into your land bank account. You then have to report to AdSense the amount deposited. After the deposit is confirmed, AdSense will send you a postcard to verify your address. You must then report to AdSense the pin number located on the postcard. Once all the verification takes place YouTube allows you to connect all of the accounts and by doing so, you can now monetize your video's and create a passive income stream.

Social Media

You should join Social Media web sites like Facebook, Google Plus, Digg, Twitter, Linkedin, Tumbler and Pinterest. Every time you upload a video. When you are finished Optimizing it, you should link it to all of your social media web sites. This creates Backlinks. A Backlink is an incoming hyperlink from one webpage to another. Google and YouTube will rank your video higher if it has a good number of Backlinks. However if you have too many, and it appears that you have created them artificially, then Google and YouTube can punish you by removing your video.

YouTube Video Marketing Overview

As long as you are backlinking organically and not using Black Hat software or Black Hat web sites, you should be find with Google and YouTube.

Show Me the Money!

Monetization involves you allowing AdSense to place ads that run before or are placed on your videos. If the ads are clicked on, you make money. If the ads are viewed in their entirety you make money.

After you have your accounts set up, you need to gather all of the tools you will be using to create videos. You can create your videos using a standard video camera and tripod and videotape yourself. Or any other number of ways you can capture video. However for this program we are going "zero cost" so there will be no need to purchase or obtain a video camera.

Getting Free Tools to Create Your Videos

We are going to use "Screen Capture" software. Go to http://screencast-o-matic.com/home to download a free screen capture software called Screencast-o-Matic. There are two versions. The Free version allows you to videotape up to 15 minutes of content and places a watermark on all of your recordings. The pro version makes longer recordings and has edit tools and not watermark. The pro version cost $15 a year and may be worth the investment once your business begins to make a profit.

YouTube Video Marketing Overview

Then next tool you will use in creating your videos is a free copy of the office software package called Apache OpenOffice. Go to https://www.openoffice.org/download/ to download the software.

100% Copyright Free Content

Now that you have to tools to create a video, you need content. Wikipedia is an excellent source of copyright free content, you can use to create your videos. There are many keyword phrases that you can use to find material. Later on in this chapter you will learn how to use the Google Ad Planner to get the best keyword phrases to use in your videos.

YouTube Video Marketing
SEO – The Key to Internet Riches

Search Engine Optimization

Analytics: Video Viewership

Through out this chapter I am going to discuss many YouTube analytics that factor into how your video is ranked in YouTube. Once someone clicks onto your video to view it, YouTube keeps track of how many minutes it was view. Videos that are viewed from beginning to end get ranked higher base on the belief that the content is good because the viewer keeps watching it. For this reason, it is usually a good idea to keep most your videos under five minutes. In addition, this allows you to create more videos to a related topic. It is better to have twenty 3 minute videos than one 1 hour video, because it is more likely that the 3 minute videos will be watched in their entirety. Also by creating 20 videos you now have 20 possible places for AdSense to place monetized ads and thus increase your earning potential 20 times.

Tags, Keywords and Keyword Phrases

Tags, keywords and keyword phrases are the most important part of getting your YouTube video to rank on the first page of YouTube. There is an old saying..."If you commit murder, where do you hide the body, where nobody will find it? On the second page of Google".

YouTube Video Marketing
SEO – The Key to Internet Riches

Although we are working on YouTube the principle is the same. You must rank on the first page of YouTube in order for your video to get views from standard YouTube web site traffic.

Keywords are words that relate to your video. Some keywords for business are:

Business, Marketing and Start-up

Keyword Phrases for business are:

how to make money from home, internet marketing, small business grants

Tags are Keywords or Keyword Phrases that you place on your YouTube video's editing page, in order to get viewers to find your video.

Your goal is to try to rank in the top 20(land on the first page of YouTube) for every or most of the Tags in your video.

Your Video Title

The title of your video should be a keyword phrase that you want to rank for. It should also be relevant to the content in the video. When your title, tags and description are all relevant it boosts your YouTube rankings.

YouTube Video Marketing
SEO – The Key to Internet Riches

Video Description

Each video is allowed to have a description. At the top of the description box, is where you should place a clickable or hyperlink, to either your web site or another video that you wish to viewer to see. Below the link should be a description of the video that contains content that is relative to the video. One short cut you can use it to cut and paste your video script into the description.

You video description should also have the keywords you used as tags. This adds to the videos relevancy.

You should also put links in you video to your social media addresses.

Half Time Adjustments

Any tags that are ranking your video in the top 20 should be placed in the headline/title of the video to boost their rank even higher.

One software that helps save you a tremendous amount of time doing this is called Tube Buddy.

https://www.tubebuddy.com/

YouTube Video Marketing
Writing Your Script

CREATING CONTENT

You have two options for creating content. On screen video of yourself using a digital camera or phone camera. Take notes of what you will discuss.

Know your topic before you hit record.

Recording Tips:

* Use good lighting.

* Try recording near a window during the day time.

* Limit background noise as much as possible.

* Use a POWERPOINT screen capture style video.

* Create bullet points

* Use free software like jing or camstudio to record it. You can also get a free 30 day trial of camtasia from TechSmith

* www.screencast-o-matic.com is another free solution.

* Use your computer's built in microphone.

YouTube Video Marketing
Writing Your Script

* Use a usb microphone is ideal, but not required.

* if you or kids have a usb gaming headset that works as well.

* most smart phones have a mp3 recording option.

Writing Your Script

Try to use words in your script that get and hold your viewers attention. Words like... you, want, now, free, limited time, All-American, imagine and how to, are just a few of the many words that are proven to stir a viewers emotions. Viewing a few copy writing videos on YouTube should help you to chose attention grabbing words.

AIDA is an acronym used in marketing and advertising that describes a common list of events that may occur when a consumer engages with an advertisement.

- A – attention (awareness): attract the attention of the customer.
- I – interest of the customer.
- D – desire: convince customers that they want and desire the product or service and that it will satisfy their needs.
- A – action: lead customers towards taking action and/or purchasing.

YouTube Video Marketing
Writing Your Script

Using a system like this gives one a general understanding of how to target a market effectively. Moving from step to step, one loses some percent of prospects.

AIDA is a historical model, rather than representing current thinking in the methods of advertising effectiveness.

A basic rule of thumb for writing your script is that one paragraph equals about 60 seconds of talking. So if you are trying to shoot a 3 minute video you what to create a 3 paragraph document for your script. Try to use words in our script that are relevant to the title of your video.

You can also cut and paste your script into a YouTube video editor, and make your video Closed Captioned. This will increase your rankings in the YouTube search engine and it will allow more people to understand your video and increase your views.

CREATING TOPICS FOR YOUR VIDEOS

It is time to brainstorm and write down topics for your videos.

Remember you could choose a video around your own information product if you had it.

YouTube Video Marketing
Writing Your Script

Get a notepad and think of 10 to 20 FAQ about your business.

http://answers.yahoo.com

Is a good source to find out what the potiential customers of your business are interested in.

Also look at articles on ezinearticles.com and see what topics come up the most for articles related to your business.

You can also browse forums related to your business.

Take a look at information products about your target market.

When you make a video that features Frequently Asked Questions each faq could be a short 1 to 3 minute video.

Use nichesuggest.com for a list of possible keyword ideas as well as seocentro and the google keyword planner.

Brainstorm 5 to 10 additional solution oriented videos. You should cover why the solution you are offering is better and why does your product recommendation solve your customer's problem.

YouTube Video Marketing
Writing Your Script

Try to think of every advantage possible. Read other reviews of similar products or businesses or view sales pages for ideas of content for your videos.

Creating a Multipurpose Close

There are certain things that you should say in almost all of your videos:

* Thank the viewer for watching

* Ask the viewer to Thumbs up or Like your video

* Ask the viewer to subscribe to your YouTube Channel

* Ask the viewer to leave a comment

* Ask the viewer to share your video link with friends or social media

YouTube Video Marketing
Writing Your Script

YOUR CALL TO ACTION

send your website visitors to a variety of places.

* A free website through weebly.com

* A free page through squidoo.com

* A free blog through blogspot.com

Use a tracking link like www.bit.ly or www.tinyurl.com

be careful as these links can change on you.

YouTube Video Marketing
Writing Your Script

UPLOADING VIDEO

Create your account at www.youtube.com you can use a google account if you have one already created. Upload your video. Then provide your keyword rich video title. Look at other examples of videos performing well in that space. Use keywords from your niche or business and topic research write a good description with the keywords in it.

Try to include at least 2 sentences in your description. More content in your description will not hurt you. Include your website link at the beginning of the description use format http://www.yourfreelink.com encourage likes, comments, or honest feedback at the end of the description. Make a call to action in the description as well.

BeeKeeping
Resources

BEEKEEPING RESOURCES

This resource section is not only a great place to get supplies to start you business, but is also shows you a few different business models to learn from. What products do they stock? Which web site design is the easiest to navigate? What prices do they charge? You don't have to reinvent the wheel. Find out what others do, and do it a little better.

The 2020 worldwide pandemic has affected many businesses. As of the release of this book, April 2021, all businesses in this section were up and running.

www.honeybeesonline.com

The Honey Bees Online is a family owned business established in 1994 with a variety of beekeeping supplies. They aslo have online training videos and a YouTube channel. So if you read the YouTube chapter of this book and were wondering if it really works...Exhibit A. It will work...if you work it. This is one of the easiest web sites to navigate and they have a great selection of different supplies for your business.

http://www.apiarybeekeepingsupplies.com/

This is a beekeeping supply web site with the greatest landing page in the history of the internet. Literally...the greatest landing page. They are based in Arkansas and have a list of local beekeepers on their website. They also have a free supply and product catalog that you can download.
870-305-1125

BEEKEEPING RESOURCES

www.beekeeping-tools.com

KingReal is a beekeeping supply factory for professional beekeeping equipment. They can customized your equipment orders. They ship worldwide and have been in business since 1995.

http://www.bbhoneyfarms.com/store/

The B & B Honeyfarm have been selling beekeeping supplies nationwide for over 40 years.
They sell beehive beauty and health aids, bee containers, beeswax, package bees, queen bees, candle supplies. They have supplies for commerical and beginner beekeepers.

www.dadant.com

Dadant beekeeping supplies. This web site has an excellent free online learning center. Free shipping with orders over $100. Started their business in 1863. They have brick and mortar offices nationwide.

www.heartlandhoney.com

Heartland Honey is based in Missouri and Kansas. They don't have a variety of supplies but they do sell honey, creamed honey and handmade soaps. This web site gives you an idea for possible business models you can have to sell your products.

BEEKEEPING RESOURCES

www.hnbeekeeping.com

Henan Beta Bee Supplier. Hena Beta specializes in the making and exporting of beekeeping equipment. They work with small and medium sized beekeeping businesses. They are based in China and export to the USA, Europe, Chile, and Africa.

www.gabees.com

Rossman Apiaries. They offer free shipping. Rossman sells package bees and Italian bees. They have beekeeper clothing and accessories. Educational products as well as extraction and packaging containers. 1-800-333-7677

https://www.modernbeekeeping.co.uk/

Modern Beekeeping is based in the United Kingdom. Their web site has a complete catolog of their products. The sell most standard beekeeping products as well as protective clothing for adults and children.

http://www.xstarpublishing.com/

This is a web site that has ebooks covering a variety of beekeeper topics.

BEEKEEPING RESOURCES

KNOW THE SCORE

SCORE, the nation's largest network of volunteer, expert business mentors, is dedicated to helping small businesses get off the ground, grow and achieve their goals. Since 1964, we have provided education and mentorship to more than 11 million entrepreneurs.

SCORE is a 501(c)(3) nonprofit organization and a resource partner of the U.S. Small Business Administration (SBA). Thanks to this generous support from the SBA and because of the selfless contributions of our more than 10,000 dedicated volunteers, we are able to deliver most of our offerings at no cost.

SCORE's Small Business Services

SCORE provides a wide range of services to established and budding business owners alike, including:

Mentoring

Entrepreneurs can access free, confidential business mentoring in person at more than 250 local chapters or remotely via email, phone and video.

BEEKEEPING RESOURCES

SCORE mentors, all experts in entrepreneurship and related fields, meet with their small business clients on an ongoing basis to provide continued advice and support.

Get connected with a mentor.

Webinars and Courses on Demand

SCORE regularly offers free online workshops on topics ranging from startup strategies to marketing and finance. Attendees can watch webinars live, or view recordings online on their own time. We also offer interactive courses on demand, so you can walk through each module at your own pace.

Register for an upcoming LIVE webinar, explore our past webinar recordings and take our courses on demand.

Library of online resources

Business owners can also benefit from SCORE's extensive collection of eguides, templates, checklists, blogs, videos, infographics and more. We strive to provide the most relevant and current educational content to help small business owners and entrepreneurs succeed.

BEEKEEPING RESOURCES

Check out our library of resources.

Local events

Many local SCORE chapters hold free or low-cost in-person workshops and roundtable discussions covering a range of topics.

https://www.score.org/

Conclusion

CONCLUSION

People join the beekeeping industry for a number of reasons, including cross-pollination of flowers, reproduction, and export purposes. Some people want to have honey and other bee goods, such as wax or propolis.

Since there is so much to discover about beekeeping, it can be a very interesting hobby for beginners. Every year, even the most experienced beekeepers come across a new situation. A basic understanding of why bees do what they do, what they need, what challenges they face, and how beekeeping tools work is required for the new beekeeper.

So if you are interested in beekeeping, always learn its basics thoroughly. Don't forget that beehives must be located in an environment where you have access to proper sunlight, as well as flowers and a water supply. It'll be best if you can find a place where bee predators can't find you.

Honeybees' activity is fully influenced by the environment in which they live. The best time to start a hive depends on your local climate and geography. To learn how others have succeeded, you need to communicate with local beekeepers and beekeeping groups in your area to learn the tricks.

One of the most enjoyable elements of beekeeping is watching the hives flourish over many seasons. This is not very simple; it requires time, persistence, and the ability to spot problems and assist your bees when they need your help.

CONCLUSION

Bees can have many problems. Some are Colony Collapse Disorder, mites, or predators. The good news is that you will find many options in this book that can help you in facing this kind of situation.

This book will provide complete information about every aspect of beekeeping. If you are a beginner or looking for any handy tool, then read this book as it will provide guidance about where to start and, most importantly, how to start.

You need to understand that just because you've done your homework, gathered all of your equipment, made the necessary arrangements, and made contact with local beekeepers doesn't guarantee that your hive will be successful every time.

While beekeeping is a rewarding hobby, it is not without its challenges, some of which may be difficult to understand.

Now that you have read this book, you are more than prepared to meet those challenges and reap all the benefits that Beekeeping has to offer!

Get excited and get started...today!

A Beginner Beekeeper's Beekeeping Book

We want to thank you for the purchase of this book and more importantly, thank you for reading it to the end. We hope your reading experience was pleasurable and that you would inform your family and friends on Facebook, Twitter or other social media.

We would like to continue to provide you with high-quality books, and that end, would you mind leaving us a review on Amazon.com?

Just use the link below, scroll down about 3/4 of the page and you will see images similar to the one below.

We are extremely grateful for your assistance.

Warm Regards, MahoneyProducts Publishing

Book Link:
https://www.amazon.com/dp/B0932CX754

Customer reviews

4.6 out of 5 stars   4.6 out of 5
6 global ratings

5 star   64%_
4 star   36%-
3 star 0% (0%)   0%
2 star 0% (0%)   0%
1 star 0% (0%)

Review this product
Share your thoughts with other customers

(Write a Customer Review)

You might also enjoy:

Crochet Business Book for Beginners
How to Start up, market, finance, stitch together your crochet or knitting small home business fortune

Imagine you can have Hassle Free All-American Lifestyle of Independence, Prosperity and Peace of Mind.

Discover how to....

* Get Free Government Grants for your Business

* Get Access to Wholesale Sources to save you Massive Money

* Learn Zero Cost Marketing for Free Advertising!

* Step by Step prepare a amazing Business Plan

* Efficiently avoid Government Red Tape

* Take Advantage of Tax Laws for your business Get $150,000 Guaranteed Loan from the SBA

* How to Incorporate to Protect Your Investment

and Much Much More!

You have the right to restore a culture of the can-do spirit and enjoy the financial security you and your family deserve. People are destroyed for lack of knowledge. For less than the cost of one night at the movies you can get the knowledge you need to start living your business dreams!

Don't wait. You'll wait your life away...

Amazon.com Book Link:
https://www.amazon.com/dp/1539606112

www.ingramcontent.com/pod-product-compliance
Lightning Source LLC
Chambersburg PA
CBHW060231030426
42335CB00014B/1406